Meaning, Mind, and Matter

Meaning, Mind, and Matter

Philosophical Essays

Ernie Lepore and Barry Loewer

OXFORD
UNIVERSITY PRESS

OXFORD
UNIVERSITY PRESS

Great Clarendon Street, Oxford OX2 6DP

Oxford University Press is a department of the University of Oxford.
It furthers the University's objective of excellence in research, scholarship,
and education by publishing worldwide in

Oxford New York

Auckland Cape Town Dar es Salaam Hong Kong Karachi
Kuala Lumpur Madrid Melbourne Mexico City Nairobi
New Delhi Shanghai Taipei Toronto

With offices in

Argentina Austria Brazil Chile Czech Republic France Greece
Guatemala Hungary Italy Japan Poland Portugal Singapore
South Korea Switzerland Thailand Turkey Ukraine Vietnam

Oxford is a registered trade mark of Oxford University Press
in the UK and in certain other countries

Published in the United States
by Oxford University Press Inc., New York

British Library Cataloguing in Publication Data
Data available

Library of Congress Cataloging in Publication Data
Data available

Typeset by SPI Publisher Services, Pondicherry, India
Printed in Great Britain
on acid-free paper by
MPG Books Group, Bodmin and King's Lynn

ISBN 978-0-19-958078-1

1 3 5 7 9 10 8 6 4 2

Contents

Acknowledgments

The authors gratefully acknowledge permission from the original publishers to reprint the following essays here.

Translational Semantics: *Synthese* 48 (1981): 121–133. © 1981 by D. Reidel Publishing Co., Dordrecht, Holland, and Boston, USA.

Three Trivial Truth Theories: *Canadian Journal of Philosophy* 13 (1983): 433–447.

What Model-Theoretic Semantics Cannot Do: First appeared in *Synthese* 54 (1983): 167–87. Reprinted with kind permission from Kluwer Academic Publishers.

The Role of "Conceptual Role Semantics": Special issue of *Notre Dame Journal of Formal Logic* on "Contemporary Perspectives in Philosophy of Language," 23/3 (1982).

Dual Aspect Semantics: *New Directions in Semantics*, ed. Ernest Lepore, Academic Press, 1986, 83–112.

What Davidson Should Have Said: *Grazer Philosophiche Studien*, ed. Rudolf Haller, Vol. 36 (1989): 65–78. Reprinted in *Information Based Semantics and Epistemology*, ed. E. Villenueva, Basil Blackwell, 1990, 190–199.

You Can Say *That* Again: *Midwest Studies in Philosophy* 14 (1989): 338–356.

Conditions on Understanding Language: *Proceedings of the Aristotelian Society*, 97 (1996, September): 41–60. (Meeting of the Aristotelian Society, held in the Senior Common Room, Birkbeck College, London, on Monday, 11th November, 1996 at 8.15 p.m.)

Solipsistic Semantics: *Midwest Studies in Philosophy* 10, (1986): 595–614.

A Putnam's Progress: *Midwest Studies in Philosophy* 12 (1988): 467–481.

Mind Matters: *The Journal of Philosophy* 84/11 (1987): 630–42. (Eighty-Fourth Annual Meeting American Philosphical Association, Eastern Division. Published by: Journal of Philosophy, Inc.)

More on Making Mind Matter: E&B *Philosophical Topics* 17/1 (1989): 175–191.

From Physics to Physicalism: *Physicalism and its Discontents*, ed. C. Gillett and B. Loewer, 2001, Cambridge: Cambridge University Press, pp. 37–56.

Mental Causation or Something Close Enough: *Contemporary Debates in the Philosophy of Mind*, ed. B. McLaughlin and J. Cohen. Malden, MA: Blackwell Publishing, 2007, 243–264.

Introduction

Many themes discussed in this book continue to ring true for us; in particular, the ideas that the correct form of a semantic theory for a natural language is a truth theory and that certain cognitive transitions are characteristic of linguistic competence. We found in Donald Davidson's work insights into these themes, and thought then—as we still do today—that others have misunderstood him, and in consequence, have failed to see why their own programs are unsatisfactory.

In recently canvassing others about Davidson's impact on semantics for natural languages, we were unsurprised to learn that most think his program has had little impact on the daily practices of semanticists. Of course, there are exceptions: virtually everyone has bought into Davidson story about action verbs quantifying over events. There is also a consensus in semantics concerning the importance of the compositionality principle; but it's unclear who gets the credit—is it Frege, Tarski, Montague, or Davidson? Further, almost every semanticist pays lip service to the Davidsonian slogan that knowledge of meaning requires knowledge of truth conditions.

In the first few chapters we outline a number of Davidsonian themes and discuss how they figure into a larger semantic project. Our first joint article on translational semantics asks: What knowledge suffices for linguistic competence? An answer requires deciding what counts as evidence of linguistic competence—which practices must we master to qualify as understanding a language?

In the 1960s and 1970s Fodor, Katz, and Postal, among others, demanded a semantics for a language that explicates the distribution of properties like synonymy, meaningfulness, anomaly, logical entailment and equivalence, redundancy and ambiguity, across all its expressions. Their semantics, sometimes called *translational* or *structural semantics*, aimed to explain or predict a set of meta-linguistic intuitions about various semantic properties of, and relations between, expressions. For example, if two expressions are synonymous, the theory ought to predict this; or if one is true as a matter of meaning alone (i.e. is analyticial), then the theory should predict this as well.

The standard criticism of translational semantics ran as follows: one could know that such structural features obtained for the expressions in a language without understanding what these expressions meant. For example, we can know that 'schneit' in German translates 'piove' in Italian without understanding either expression. So, if knowledge of a semantic theory must suffice for understanding its object language, then more is required than is provided by a translational semantics.

Our account of the semantic project centreed not on the distribution of these meta-linguistic properties and relations, but on certain transitions from linguistic to non-linguistic beliefs. Suppose someone you trust affirms in language L a sentence S; and suppose further on this basis that you (but not someone who does not understand L) can come to believe (or are justified in believing) that *p*. Knowledge of what sort of information would suffice to account for these transitions? For example, suppose A utters 'Es schneit' in a language you understand; if you trust A and find her reliable, you would be justified, *ceteris paribus*, in believing that it's raining. What might you know about 'Es schneit' to account for this transition?

Our early chapters argue that any theory that fails to account for such transitional aspects is incomplete; and further that an adequate explanation requires knowledge of truth conditions. For example, knowing that 'Es schneit' is true in German just in case it's raining, would (partially) explain why the transition from a heard utterance of 'Es schneit' to the belief that it is raining is justified. We also pressed the productivity of these transitions—there is no upper limit on how many a linguistically competent speaker is justified in making—to defend positing the goal of devising a compositional semantic theory, one, according to us, that issues in correct interpretive truth conditions for each (indicative) sentence. In short, we defended absolute truth theories à la Tarski–Davidson as having the appropriate form for a semantic theory to take.

Though most theorists agreed on the shortcomings of translational semantics, few thought linguistic competence could be explicated by invoking a Tarski-like absolute truth theory. For, as Alonzo Church once observed, even though a truth theory that meets Tarski's Convention T—that is, one that ensures that the sentence mentioned on the left hand side of a T-sentence translates (or interprets) the one used on the right hand side—gives the meanings for its object language sentences, it does not say that it's so doing. The theory asserts no more than the extensional equivalence of the sentences mentioned and used. But then, so the argument goes, the truth theory comes unhinged because we demand that it enable us to determine what others say when they speak. And if all we know were conditions under which his words were true, for example, that ' "Es schneit" is true iff it's raining', while this may justify our coming to believe that it's raining, it does not justify our concluding that he said it's raining when he utters the German sentence. In effect, the charge requires of a semantic theory for a natural language that it enable its knower to determine what speakers say when they speak.

We expended much energy arguing that the competing semantic frameworks fail; for example, Lepore argued that Model Theoretic Semantics (MTS), surprisingly, is no advance over translational semantics. For, while proponents of MTS rejected translational semantics on the grounds that it failed to capture the appropriate word-world relations, MTS also fails to engineer the relevant transitions between language and world. This is because MTS never directly tells us the truth conditions or meanings of sentences. Instead, it issues truth conditions for sentences *relative to an interpretation* (a model). But we cannot derive an absolute truth theory from a relativized truth theory, and it is only with an absolute truth theory that we can figure out what the world is like, given the presumed truth of what we hear. In particular, we cannot justifiably infer that it's raining from the fact that our reliable friend utters 'Es regnet' if we only know a relative truth theory. To derive an absolute truth theory from a MTS, however, we would need to add either an absolute truth theory or enough knowledge to distinguish the actual world from all other worlds. Needless to say, we lack such knowledge!

In the 1980s, mostly in reaction to arguments for externalism from Putnam, Kaplan, Kripke, and others, many authors turned to Dual Aspect Semantic theories (DAT)—one tier corresponding to a truth theory, the other to a conceptual role theory. To say S means that p on this account requires not only that S and p share truth conditions but also conceptual roles. (We hasten to add that the conception of truth conditions in this literature typically differs from the one in the Davidson program.) We criticized DAT by arguing that the claim that two expressions share meaning only if they share conceptual role was far too restrictive. We speculated then that perhaps a robust notion of a similar conceptual role might avoid our criticism. But later, in the 1990s, Fodor and Lepore (1992, 2002: in both *Holism* and *The Compositionality Papers*) argued against this possibility as well and against the very utility of a conceptual role in general.

One might wonder why we simply did not embrace a theory about meaning directly instead of all this indirection through truth, models, conceptual roles, and translation. In our contributions, we tended to minimize the need for invoking meaning theories by arguing that truth theories alone would suffice. But suppose it turned out that truth theories alone were too weak to explicate linguistic competence. Would it not then be appropriate to invoke a meaning theory directly? We noted in response to this question, following Davidson, that a meaning theory might be either a theory that quantifies over meanings and interprets expressions of a language by assigning these meanings to them; or, instead, it might be no more than a theory that instead of issuing in T-sentences that ascribe truth conditions (without reifying them), issues in M-sentences that ascribe meaning to sentences (without reifying them), as in, '"Es schneit" means that it's snowing'. One might even argue—as many did—that to aim for anything weaker is to miss the point of semantics.

Davidson expressed reservations about the possibility of devising meaning theories in either sense. We, in several of the chapters that follow, concur with Davidson, even suggesting (incorrectly, in retrospect) that M-clauses for sub-sentential components are ill-formed. We also expressed concern over how to identify principles we could invoke in deriving the infinitude of M-sentences. Later, Lepore in collaboration with Ludwig (2005) pressed Davidson's Third Man Argument against meanings altogether (if by a meaning theory one means a theory that posits meanings as entities, and interpreting expressions of a language by assigning these entities to each expression).

Another charge against Davidson, as noted above, was that truth theories are too weak to issue in meanings. The concept of meaning is two-sided. On the one hand, it connects to truth, as in: if S is true and S means *p*, then *p*. But, it also connects to a host of intentional concepts, as in: if A *asserts* S and S means *p*, then S asserts *p*. This duality permits those who know the meaning of an expression to move in either of two directions when confronted by its utterance: we can infer something about the world or about the speaker himself. Though a truth theory fares well on explicating word–world relations, it seems mute as regards word–mind relations.

The standard ways of trying to accommodate the weakness of a truth theory are all well-known to fail—for example, invoking the nomologicity of its axioms and theorems; or ensuring that the theory is confirmed under radical interpretation (cf. Fodor and Lepore 1990 and Lepore and Ludwig 2005). However, in 'What Davidson Should Have Said', we argue that Davidson can avoid the charge of weakness by exploiting his paratactic account of indirect discourse. In short, we argue that both the critical literature and Davidson's response to its charge that truth theories are too weak are misguided. Our solution to the weakness charge derives from recognizing the role of indirect reports—a theme Cappelen and Lepore (2005) pursued later.

So, what if anything has changed in our minds since these collaborations in the 1980s? Lepore now thinks absolute truth theories must be constrained to be interpretive to register the right results (Lepore and Ludwig 2005), but both of us still advocate a truth theoretic approach.

On the critical side, neither Fodor nor Schiffer were impressed by the attacks on translational semantics. They pressed back with a *tu quoque*, claiming truth theories were no better positioned than translational semantics or conceptual role semantics. In response, we charged use/mention confusion, or change of topic. We granted that a truth theory does not, and indeed cannot, reveal what our grasp on the contents of our thoughts consists in. As Loewer notes (Chapter 4), this would involve a regress since we would have to explain knowledge of the truth conditions of a language of thought by postulating another language of thought understood by the knower. Furthermore as Lepore (Chapter 8) notes, the truth theory only pushes this question of in virtue of what do our words mean what they mean back to the meta-language—back to the language of thought if you like. But our critical point still stands: a translational or conceptual role semantics doesn't help, but a truth theory does. We were surely right about this, but if your goal is the further metaphysical one then it is true that the truth theory is no better suited than the translational semantics.

Our contributions 'Solipsistic Semantics' (Chapter 9) and 'Putnam's Progress' (Chapter 10) concern issues in the metaphysics and epistemology of meaning. Until the 1970s, it was widely supposed that the contents of one's thoughts and beliefs, are determined entirely internally. Frege, for example, seems to have thought that whether a person grasps a particular sense or thought entirely depends on her mind and its relation to the thought. Since senses and thoughts are abstract necessary entities, another way of putting this point is that the content of a person's thoughts *supervene* on her mind. Naturalistically minded philosophers expressed this by saying that the contents of a person's thoughts supervene on properties intrinsic to her brain. Kripke, Putnam, and Burge led a revolution that rejected this. Putnam's famous 'Twin Earth' thought experiments persuaded most philosophers that thought content is determined not only by intrinsic brain properties but also by elements of the thinkers environment, especially causal relations between her mental states, environment, and her relations to other members of her linguistic community. Davidson had already held a theory of the determinants of meaning—his theory of radical interpretation—that had this consequence. In 'Solipsistic Semantics', we pushed this idea further than it had been pushed before and argued that practically all our ordinary concepts have their contents externally individuated. Externalism of this extreme sort has seemed to many to threaten first person epistemological authority to the contents of one's thoughts. If what a person thinks is constituted not only by what is in his brain but by his causal relations to his environment, then how can he know what he is thinking? In 'Solipsistic Semantics' we suggested a kind of deflationary sense in which people can have first person knowledge of the contents of their thoughts even if externalism is true.

Putnam famously argued that if what he called 'metaphysical realism' is true, then it will be impossible to provide a naturalistic account of how our words and thoughts come to have contents. Putnam characterizes metaphysical realism in various ways but its core commitment is that the properties, entities, and facts in the world exist completely independently of our conceptualizations of them. Somehow this independence rules out our thoughts being about such properties, entities, and facts. Putnam contrasts this with what he calls 'internal realism', which that holds that our conceptualizations are in some way involved in constituting the properties, etc. that we think about. Most of our chapter is devoted to an attempt to clarify Putnam's argument against metaphysical realism. In the end, we conclude that if there is a problem about there being a naturalistic account of the content of thoughts it is as much a problem for internal realism as it is for metaphysical realism.

The last four chapters in this collection discuss Davidson's account of the relation between the mental and the physical and the problem of mental causation.

In his seminal paper 'Mental Events,' Davidson formulated a version of the mind–body problem in terms of this apparently inconsistent triad:

(1) Causal interaction: at least some mental events interact causally with physical events. Mental events can cause physical events, e.g. my thought that it is raining causes me to grab an umbrella.

(2) Nomological account of causation: events related as cause and effect are subsumed under a strict law.

(3) Anomalousness of the mental: there are no strict laws formulated in mental vocabulary, and specifically no strict laws specifying physical properties as sufficient for mental properties.

If causation requires subsumption under strict laws and there are no strict laws connecting mental and physical predicates/properties, then how can mental events cause or be caused by physical events? Davidson's solution is that mental events are identical to physical events. For example, that Don's noticing at t that he is late is identical to some sequence of neuron firings in Don's brain. Indeed (1), (2), and (3) *entail* the token identity of mental events. On the other hand, (3) apparently entails that mental properties are distinct from physical properties. So if a mental event m causes a physical event p then m is identical to a physical event r. That is, m has a physical description P (or satisfies a physical property) and this physical description is connected by strict law to a physical description of p.

Anomalous Monism (AM) seemed at first to provide a happy reconciliation of physicalism with the apparent anomalousness of the mental. However, it didn't take long for philosophers to start worrying that AM was really a version of epiphenomenalism, what some called 'property epiphenomenalism.' The problem is that although AM endorses mental event causation, it seems to make mental properties irrelevant to those causal relations since it is *physical* properties that occur in strict laws and ground causal relations. The point was sometimes put by claiming that if AM is true then the counterfactual (MR) 'if r had not been M then p would not have occurred (or r would not have caused p)' is false. But this seems to express the causal irrelevance of M. On the widely held Goodmanian account of counterfactuals, a counterfactual is true only if the antecedent, truths cotenable with the antecedent and laws, logically imply the consequent. Since according to AM there are no strict mental laws it seems to follow that there are no mental-physical counterfactuals of the sort that expressed the causal relevance of mental properties.

In 'Making Mind Matter' and 'More on Making Mind Matter,' we pointed out that the view that (MR) is incompatible with (AM) depended on this Goodmanian account about counterfactuals and that this account is not plausible and had been superseded by David Lewis's similarity account. On Lewis's account, while laws still play a special role in determining the world similarity relation, there is no requirement that there be laws, let alone *strict* laws, connecting the mental with the physical for counterfactuals like (MR) to be true. We argued that given Lewis's account, not only are counterfactuals like (MR) compatible with (AM), but more complex ones like (MR★) "if r had occurred and been M but not P then p would still have occurred (or would still have been caused by r)" may be true. The idea is that in the most similar worlds in which r occurs and is M, it is also P★ and p still occurs (and is still caused by r). This seemed to us to successfully deflect the objection that (AM) renders mental properties causally impotent. What more could one want from mental causation than that the physical situation (including one's bodily movements) counterfactually depends on the mental properties of mental events.

(AM) is a version of 'non-reductive physicalism' (NRP). It holds that physics is nomologically closed and every event is a physical event (the physicalist part). AM also says that mental properties are distinct from physical properties (the non-reductive part). So far this is a rather weak kind of physicalism since it allows for mental properties to float freely from physical properties. To strengthen it, Davidson added to AM the claim that mental predicates/properties *supervene* on physical predicates/properties. But exactly how should 'supervenience' be understood? One construal—so called 'Weak Supervenience' (WS)—says that within any possible world, if two events differ in their mental properties they differ in physical properties. A stronger construal—'Strong Supervenience' (SS)—says that for any pair of events, whether they are world mates or not, if they differ in mental properties, then they differ in physical properties. This implies that for mental property M there is a physical property P such that

P is nomologically sufficient for M. Perhaps Davidson was cagey about choosing between WS an SS because he rejected talk of 'possible worlds'.

(SS), or its near relative 'global supervenience' (GS)—the claim that any world that is a minimal physical duplicate of the actual world is a duplicate *simpliciter*—better captures the physicalist idea that the totality of physical facts determines the totality of mental facts. But why should one accept (SS)? In 'From Physics to Physicalism', Loewer argues from the causal relevance of the mental (understood in terms of counterfactuals construed along Lewisian lines) to (GS). While (GS) does not require the existence of strict laws (as Davidson understands them) connecting physical with mental properties or that mental properties are identical to physical properties it does apparently violate Davidson's claim that 'there is no tight connection' between the mental and physical.

According to Jaegwon Kim, neither Davidson's nor any other version of NRP can accommodate mental causation. He argues that if physics is causally/nomologically closed then either there is no mental to physical causation or mental properties are identical to physical properties. This is his famous 'exclusion argument'. Loewer's 'Mental Causation or Something Close Enough' argues, contra Kim, that if causation is understood in terms of counterfactuals then Kim's argument fails. So the account of mental causation in terms of counterfactuals is able to vindicate a version of NRP, albeit a version that may involve a stronger physicalism than AM.

References

Cappelen, H. and E. Lepore (2005) *Insensitive Semantics*, Oxford: Basil Blackwell, 2005.

Fodor, J. and E. Lepore (1992) *Holism: A Shopper's Guide*, Oxford: Basil Blackwell.

—— and —— (2002) *The Compositionality Papers*, Oxford: Oxford University Press.

Lepore, E. and K. Ludwig (2005) *Donald Davidson: Meaning, Truth, Language, Reality*, Oxford University Press.

—— and —— (2007) *Donald Davidson's Truth-Theoretic Semantics*, Oxford: Oxford University Press.

1

Translational Semantics

Ernest Lepore

David Lewis has criticized semantic theories which assign meanings to sentences of a language L by translating them into some other language L⋆. Even if L⋆ has been expressly designed to exhibit particular semantic features, he thinks such a theory would nevertheless fail to be a theory of meaning for L. His primary target is Katz and Postal's theory—which, he claims, interprets English by translating it into a symbolic language he calls 'Semantic Markerese'. The objection is that

we can know the Markerese translation of an English sentence without knowing the first thing about the meaning of the English sentence: namely, the conditions under which it would be true. Semantics with no truth conditions is no semantics. (Lewis, 1972, pp. 69–70)[1]

Clearly, Lewis thinks that an adequate semantics for L must assign to the (indicative) sentences of L their truth conditions and that translational semantics fails to do this.

Jerry Fodor and Gilbert Harman have each responded to Lewis' criticism by arguing that he overstates the difference between translational semantics and truth-conditional semantics (Fodor, 1975; Harman, 1971, 1974). Fodor parries the criticism by arguing that Lewis' objection applies equally to truth-conditional theories. Harman agrees and further claims to have shown how to convert a translational semantics into an 'equivalent' truth-conditional semantics. He also argues that truth conditions have less to do with meaning than is commonly supposed, and that truth-conditional semantics illuminates meaning not by assigning truth conditions but by exhibiting the role of logical words such as 'and' or 'or' in its recursion clauses.

A number of other authors have agreed with Harman and Fodor in seeing little difference, if any, between specifying meaning by translation and specifying meaning in terms of truth conditions.[2] In this chapter I argue that all these authors have misunderstood Lewis' objection. I will show that, at least with respect to one task that semantics is reasonably called on to perform, there is an important difference

---

[1] By a translation semantic theory for L we mean a theory which entails, for each sentence S′ of L, a sentence of the form "S in L translates as S⋆ in L⋆." A truth condition semantic theory for L entails for each S a sentence of the form "S is true iff *p*."

[2] See for instance Field, 1977.

between a truth-conditional semantics and a translational semantics. By focusing on this task I not only explain why translational semantics is inadequate, but also provide a partial justification for the slogan that to give the meaning of a sentence is to give its truth conditions.

What should a semantics for a language L accomplish? There is no uniquely correct answer to this question. Here are some: it should show how the meanings of complex expressions depend on the meanings of their constituents; it should account for logical and other semantic features of a language (like logical consequence, logical truth, synonymy, analyticity, and so on); it should provide an account of how the illocutionary force of an utterance is determined on the basis of its semantical features and context. These are worthy undertakings. Here I will focus on a related task, but a relatively simple one. Consider the following—simple—communication episode:

Arabella looks out the window, turns toward Barbarella and asserts in their common language L these words: 'Es schneit.' Barbarella hears these words and, on that basis, justifiably acquires the belief that it is snowing.[3]

Suppose that someone, we will call him Interpreter, witnesses the scene described. Further, suppose that later on he discovers that Barbarella, upon hearing Arabella utter 'Es schneit' came to believe, justifiably, that it is snowing. However, he was not himself justified in this belief upon hearing Arabella's utterance, because he does not understand L.

A natural way of characterizing the difference between Barbarella and Interpreter is that she has knowledge through understanding L, which he lacks. If it is propositional knowledge that distinguishes her from him and characterizes her understanding of L, it must be enormously complicated. In our episode, it is plausible for instance that Barbarella knows that 'Es schneit' is a sentence of L, that Arabella asserted that it is snowing, that her utterance means that it is snowing, and so on. For the purpose of this chapter we will assume that Barbarella's understanding of L consists, at least in part, of her having certain propositional knowledge. We want to focus in particular on one part—an especially significant one: the knowledge that warrants Barbarella's belief that it is snowing on the basis of her belief that Arabella asserted the words 'Es schniet.'

In our scenario we isolated one of Barbarella's beliefs as requiring justification and we neglected any other she may acquire. But, of course, she may acquire beliefs other than that it is snowing on the basis of Arabella's utterance—for example that it is cold outside, that the university will close, and so on. Interpreter will not know beforehand which of these belief acquisitions needs to be justified by citing her non-semantic as well as her semantic knowledge. For our purposes, there need not be a sharp distinction between these two sorts of knowledge. No matter how we make the distinction (or even if we do not make it), it appears that some knowledge about their common

---

[3] We use the expression 'utters' in the present context (in contrast to 'says that') in such a way that a speaker may utter (on a particular occasion) some words without her or our knowing what these words mean.

language will be required in order to justify many of the indefinite number of beliefs Barbarella may acquire on the basis of Arabella's utterances. We begin by supposing that Barbarella justifiably believes that Arabella is a speaker of L and is, in this instance at least, reliable (in other words whatever Arabella asserts is usually true). On the basis of this and of her belief that Arabella asserted 'Es schneit' Barbarella is justified in believing that 'Es schneit' is true. What additional knowledge would justify her concluding that it is snowing?

A short answer is that her knowing the meaning of 'Es schneit' justifies her believing that it is snowing. But this answer, although perhaps correct, is unilluminating in this context, since it does not specify the meaning of 'Es schneit' in such a way that Interpreter can discern it. If Interpreter already understood Arabella's language, then the short answer would explain to him why Barbarella is justified in believing that it is snowing. But we have assumed that he does not understand L. Knowing our short answer puts him in no better position to explain why Barbarella is justified in interpreting any particular sentence—say, 'Es schneit'—as 'It is snowing'.[4] How, then, should Barbarella's knowledge be specified?

Katz and Postal's translational semantics suggests one kind of specification. For our purposes it is not important to discuss in any detail how their theory develops. Suffice it to say that Lewis thinks semantic interpretation within this theory ultimately involves translating natural language sentences into sequences of objects called 'semantic markers' (Katz and Postal, 1964).[5] That is to say, their theory ultimately issues in consequences which express translations between an expression of a natural language and an expression of Semantic Markerese. The kind of specification of Barbarella's knowledge that this account suggests is that she knows that 'Es schneit' translates as M (where M is a sentence of Semantic Markerese). But Barbarella's knowing this by itself does nothing to justify her belief that it is snowing. To see that this is so, note that the following inference is invalid:

'Es schneit' is true.
'Es schneit' translates as M.
It is snowing.

---

[4] Since Interpreter presumably understands some language, his knowing that the sounds Arabella makes have meaning and that Barberella knows these meanings certainly provides him with some understanding of why Barberella is justified in acquiring beliefs upon hearing Arabella's utterance. But this understanding does not derive from his knowledge of our short answer alone. This knowledge, together with his prior knowledge about understanding a language, provides him with whatever insight he has into Barbarella's justification. But, since this is so, it is inappropriate to exploit it in describing his knowledge.

[5] These translations are constrained (this is the real reason for bringing them in at all, together with the markers) such that expressions in the natural language are mapped onto the same ordered sequence of semantic markers, anomalous expressions are not mapped onto any sequence of semantic markers, ambiguous expressions are mapped onto different sequences and so on. Taken together, these semantic properties and relations provide a reasonably good initial conception of the subject matter of semantics. But our question is whether the theory devised to account for them can accomodate the range of semantic facts we are concerned with.

Clearly, 'it is not snowing' is logically compatible with these premises. It is perfectly possible for the words uttered by Arabella, 'Es schneit' to be true and for it not to be snowing. Of course, such a possibility would be one in which 'Es schneit' has a meaning different from the meaning it has. We cannot assume that it has the meaning it has, since it is the meaning of 'Es schneit' that we are attempting to specify.

The trouble with specifying meaning by translation is that knowing that 'Es schneit' translates as M does not warrant the belief that it is snowing on the basis of the belief that 'Es schneit' is true. We think it is precisely this point that is the crux of Lewis' objection to translational semantics.

If knowing that 'Es schneit' translates as M does not suffice to justify Barbarella's belief that it is snowing, what other piece of knowledge will? Suppose we attribute knowledge to Barbarella that 'Es schneit' means that it is snowing. The reasoning that would justify her believing that it is snowing from the premise '"Es schneit" is true' is now clear:

(1) 'Es schneit' is true.
(2) 'Es schneit' means that it is snowing.
(3) If a sentence S means that _p_ and S is true, then _p_.
(4) It is snowing.

The third premise bridges the gap between meaning and truth; without it both these concepts lie idle. Given that Barbarella knows (or believes) these premises, she is justified in believing that it is snowing. Note, however, that something less than her knowing that (2) and (3) are the case would suffice to justify her belief that it is snowing. If (2) and (3) are replaced by:

(2′) 'Es schneit' is true iff it is snowing

then the resulting argument is still valid.

Replacing (2) and (3) by (2′) might be thought to be advantageous at least by those who consider truth to be a clearer semantical concept than meaning. For our purposes in this chapter, it is enough to observe that both (1)–(3) and (1)–(2′) provide justification for (4). Whether or not knowledge of truth conditions is all there is to knowing, the meaning of a sentence is a question we will return to at the end of this chapter.

The question: what knowledge justifies Barbarella's belief that it is snowing on the basis of her hearing Arabella utterance 'Es schneit'? has served as a magnifying glass to make obvious the difference between translational and truth-condition semantics. I now want to diagnose Fodor' and Harman's responses to Lewis.

Fodor and Harman seem to claim that Lewis' objection that one can know the Markerese translation of an English sentence without knowing the first thing about its meaning applies equally to Lewis' own truth condition account:

This will hold for absolutely any semantic theory whatever so long as it is formulated in a symbolic system, and of course, there is no alternative to so formulating one's theory. We're _all_ in

Sweeney's boat; we've all gotta use words when we talk. Since words are not, as it were, self-illuminating like globes on a Christmas tree, there is no way in which a semantic theory can guarantee that a given individual will find its formulas intelligible.

So, the sense in which we can 'know the Markerese translation of an English sentence without knowing . . . the conditions under which it would be true' is pretty uninteresting.[6]

[T]here is a sense in which a theory that would explain meaning in terms of truth conditions would be open to Lewis' objection to Katz and Postal's theory of semantics markers. Lewis says, you will recall, 'But we can know the Markerese translation of an English sentence without knowing the first thing about the meaning of the English sentence: namely, the conditions under which it would be true.' Similarly, there is a sense in which we can know the truth conditions of an English sentence without knowing the first thing about the meaning of the English sentence. To borrow David Wiggins' example, we might know that the sentence 'All mimsy were the borogroves' is true if and only if all mimsy were the borogroves. However, in knowing this we would not know the first thing about the meaning of the sentence, 'All mimsy were the borogroves.' (Fodor, 1975, pp. 120–121)

Apparently Fodor and Harman construe Lewis as saying that one must understand the language in which the canonical representation is expressed before one can use the semantic theory to determine what the represented sentence means—and this is a problem any semantic theory must face. For example, one must understand the English sentence

'Es schneit' translates 'it is snowing'

if this sentence is to provide one with an account of the meaning of 'Es schneit.' Similarly, one must understand

'Es schneit' is true iff it is snowing

If this sentence is to provide an account of the meaning of 'Es schneit' If knowledge is stated in a language which Interpreter does not understand, then that statement of the piece of knowledge in question is *useless* to him.

This is certainly correct, but Lewis' point is not this obvious one; instead Lewis is arguing that someone who understands a translation and knows it to be true need not

---

[6] A reason for not making the replacement is that, while knowledge of (2′) will suffice for justifying Barberella's belief that it is snowing, this knowledge is not sufficient to justify other beliefs that she apparently acquires on the basis of her understanding of L. For example, Barberella can also come to believe that Arabella believes that it is snowing, or that Arabella said that it is snowing. To be justified in concluding that Arabella said that it is snowing, it appears that a premise like ' "Es schneit" means that it is snowing' is needed. The inference would go like this:

Arabella uttered 'Es schneit'.
'Es schneit' means that it is snowing.
If someone utters S assertively and sincerely, and S means that p, then that person says that p.
Arabella said that it is snowing.

The corresponding inference with truth is obviously false:
If someone utters S assertively and sincerely and S is true if and only if p, then she says that p.
'$2 + 2 = 4$' is true if and only if snow is white. But in uttering '$2 + 2 = 4$' I say nothing about snow being white.

know the meaning of the sentence of the translated language. It is just this that we established when we argued that attributing to Barbarella the knowledge *that* 'Es schneit' translates as M does nothing to justify her belief that it is snowing.

Because a number of writers have been misled on this point, we will perhaps be forgiven for belaboring it. No one denies that I cannot understand sentence (5) unless I understand English:

(5) The sentence 'Es schneit' in German translates the sentence 'it is snowing' in English.

Similarly, no one denies that I cannot understand sentences (6) and (7) unless I understand English:

(6) The sentence 'Es schneit' is true in German if and only if it is snowing.
(7) The sentence 'Es schniet' in German means that it is snowing.

But, whereas knowledge that (5) (together with knowledge that (1) and (3) alone) does not justify Barbarella's belief that it is snowing, knowledge of (6) or (7) does. In part, this is because knowledge of (6) or (7) does not require any competence with English. Simply note that, whereas (5) to (7) are all grammatical, the following sentence is not:

The sentence 'Es schneit' in German means that it is snowing in English.

I need not know any more English to know that (6) or (7) is the case than Galileo knew for us to say of him, truthfully, that he knew the earth is round. Moreover, as we have seen, knowledge of (6) or (7), contrary to knowledge of (5) helps to complete the chains of reasoning we have been probing.

Fodor also claims that 'there is a sense in which we can know the truth conditions of an English sentence without knowing the first thing about the meaning of the English sentence' (Fodor, 1975 p 121.) The example he borrows from Wiggins to show this is peculiar. Does he really mean that someone might know that 'All mimsy were the borogroves' is true if and only if all mimsy were the borogroves? His example makes sense only in case '"All mimsy were the borogroves" is true if and only if all mimsy were the borogroves' is a sentence in English (or extended English). Otherwise the sentence which results from prefixing to it 'Someone knows that' would be nonsense. But if it is a sentence in English, then someone who knows what it expresses will be justified in believing that all mimsy were the borogroves when he hears 'All mimsy were the borogroves' asserted by a reliable English speaker.

It is my guess that Fodor confused this sentence with the following one:

(8) 'All mimsy were the borogrove' is true if and only if 'all mimsy were the borogroves' is true.

Of course, someone might know that (8) is the case without knowing much about the meaning of 'All mimsy were the borogroves'. But then this knowledge does not attribute him knowledge of truth conditions, but the quite different knowledge that a certain English sentence is true.

Harman also has an argument which is supposed to establish that truth-condition semantics and translational semantics are 'equivalent':

It is easy to see that a theory of meaning in this sense is equivalent to a formal theory of translation. Suppose that we have a formal procedure for translating a language *L* into *our* language. Suppose in particular that we have a recursive procedure for recognizing the relevant instances of '*s* (in *L*) translates into our language as *t*.' Then we can easily formulate a recursive procedure for recognizing relevant instances of '*s* (in *L*) means *p*' or '*s* (in *L*) is true if and only if *p*' (where what replaces '*s*' is the same name of a sentence as what replaces '*s*' in the previous schema and what replaces '*p*' is the sentence named by what replaces '*t*' in the previous schema). Then we treat each of the instances of one of the latter schemas as axioms in a formal theory of truth, since each of the infinitely many axioms in the theory will be formally specifiable and recognizable. Similarly, given a formal theory of truth or a formal theory of meaning in this sense, we can easily state a formal theory of translation. (Harman, 1974, p. 6; my emphasis)[7]

Harman's procedure for constructing a truth (or meaning) theory from a translation theory is the following. Suppose that TR is a translation theory which, for each sentence *s* of L, has a consequence of the form:

    s (in L) translates as *t* (in L⋆).

Harman assumes that L⋆ is *our* language, say English. He observes that anyone who understands English will, for each sentence *t*, know the truth of a sentence of the form:

    t is true if and only if *p*,

where '*p*' is the sentence named by *t*. The truth theory Harman constructs, T⋆, results from adding to TR each instance of '*t* is true if and only if *p*' (Harman, 1974, p. 6).

We first observe that TR and T⋆ are not equivalent as theories of meaning. Suppose that TR entails:

    (9) 'Es schneit' translates as 'it is snowing.'

Then the corresponding T⋆ will entail:

    (10) 'Es schneit' is true iff it is snowing.

We will assume that (9) implies:

    (11) 'Es schneit is true' iff 'It is snowing' is true.

Then Harman's claim, restated, is that (10) and (11) are necessarily equivalent. I have already argued that the knowledge expressed by (10) will, but the knowledge expressed by (11) will not, suffice to justify the belief that it is snowing on the basis of the belief

---

[7] Cf. also Harman, 1971, p. 74.

that 'It is snowing' is true given that 'Es schneit' is true. Knowledge of what (11) expresses does justify the belief that 'It is snowing' is true given that 'Es schneit' is true; but that is quite another matter. These considerations are enough to show that a truth theory and the corresponding translation theory are not equivalent as theories of meaning. Using possible world machinery, we can also see that (10) and (11) are not necessarily equivalent. Sentence (11)—and also sentence (9)—is true at a world w at which 'Es schneit' and 'It is snowing' have the same meaning even if they happen to mean that the sun is shining. If the sun is shining at w, then (11) is true at w but (10) is false at w, since '"Es schneit" is true' is true at w but 'It is snowing' is false at w. Of course, if we restrict attention to worlds at which 'Es schneit' and 'It is snowing' are given their standard interpretation (that is, as being true iff it is snowing) then (10) and (11) are equivalent with respect to those worlds. But a theory of meaning is supposed to specify the interpretation of the sentences of a language. By restricting attention to worlds in which sentences have their standard interpretations we are, in effect, presupposing instead of giving that specification.

Although (10) and (11) are not equivalent, it is not difficult to guess why Harman thinks that they are. If

(12) 'It is snowing' is true iff it is snowing

were a necessary truth, then (10) and (11) would be necessarily equivalent. Now it is very tempting to think that (12) is necessarily true. In defense of this position, it might be pointed out that (12) is a logical truth since it is true in virtue of its form. Any sentence of the form '"S' is true iff S' is true in virtue of the disquotational effect of the truth predicate. Even if (12) is a logical truth, it does not follow that it is necessarily true. Its status is like that of 'I am here now'—a sentence which is arguably a logical truth, but which expresses a contingent truth. It is clear that (12) does not express a necessary truth, since there are possible worlds in which the words 'It's snowing', uttered by Arabella, are true iff it's not snowing. In response to this it might be replied that '"It's snowing" is true in English iff it's snowing' is a necessary (as well as logical) truth, since in any possible world in which 'It's snowing' has truth conditions other than that it's snowing this is not a sentence of English (but of some other language).

While this manoeuvre may render a sentence which states truth conditions a trivial truth, it is now far from trivial that a particular utterance of 'It is snowing' is a sentence of English. It is a sentence of English iff the utterance is true iff it's snowing. So, for Arabella to know that a particular utterance of 'It's snowing' is true in English, she would have to know (or it would have to follow logically from what she knows) that 'It's snowing' is true iff it's snowing; and that, I have argued, is not at all trivial.

Our arguments have shown that a theory which specifies the truth conditions of the sentences of a language can serve to characterize part of the knowledge involved in understanding a language, while a theory which specifies translation from one language to another cannot accomplish the same task. How far will a theory of truth take us toward specifying the meanings of a language? Further than Harman thinks, but perhaps not as far as some philosophers, for instance Davidson, have hoped. Further if our argument is correct, then to know the meaning of a sentence of L involves being able to make the inference from the truth of the sentence to its truth conditions. There are other inferences that Barbarella can justifiably make from her belief that Arabella uttered 'Es schneit'. For example, she could naturally conclude that Arabella believes that it is snowing, perhaps using an argument like this:

(13) Arabella utters 'Es schneit'
(14) If Arabella utters a sentence S then she believes that S is true

so

(15) Arabella believes that 'Es schneit' is true
(16) Arabella believes that 'Es schneit' is true iff it's snowing

so

(17) Arabella believes that it's snowing.

The key step in this argument is (16). An interpreter who attributed knowledge of what (16) expresses to Barbarella would be in a position to understand why she acquired the belief that Arabella believes that it is snowing after the latter uttered 'Es schneit'. It would be very nice if all of Barbarella's semantic knowledge of L were expressible as knowledge of truth conditions or as knowledge that other users of L know the truth conditions of the sentences of L (or know that users of L know that users of L know the truth conditions of the sentences of L). But this does not appear to be so. For example, Barbarella would be employing semantic knowledge to conclude from (13) that

(18) Arabella asserted that it is snowing.

By employing what has been called 'a theory of force' (or rather by employing knowledge of certain consequences of a theory of force), Barbarella could conclude that Arabella's utterance is an assertion. But how could she go from this conclusion to (18)? Knowledge that 'Es schneit' is true iff 'It's snowing' is insufficient, since Barbarella may also know that 'It's snowing' is true iff the temperature is below thirty degrees (assuming, say, that it's snowing and the temperature is below thirty degrees on the occasion of the utterance), but she would not be entitled to conclude that Arabella asserted that the temperature is below thirty degrees. (Of course she would be entitled to conclude that the temperature is below thirty degrees. But there would be nothing wrong with her drawing this conclusion on the basis of her beliefs.) For similar reasons, her belief that Arabella believes that 'Es schneit' is true iff it's snowing does not entitle her to conclude (18). While we have no proof of this claim, what seems to be required is Barbarella's knowing that 'Es schneit' (as uttered by Arabella on this occasion) *means that* it's snowing, and her also knowing that, if someone produces an utterance S which has the force of an assertion and the utterance means that *p*, then that person has asserted that *p*. But this knowledge goes beyond knowledge of truth conditions and involves knowledge of meanings. Whether or not it is possible to construct a theory of meaning for a language L (a theory which entails, for each sentence S of L, a true sentence of the form 'S means that *p*') is an interesting and difficult question, which we leave unanswered here.

Acknowledgments

A version of this chapter was presented to the 1980 Pacific Americal Philosophical Association meetings in San Francisco. I would like to thank Louise Anthony, Richard Grandy, Merrill Hintikka, and Esa Saarinen for helpful suggestions on earlier versions of this chapterr.

References

Field, H. (1977) Logic, Meaning and Conceptual Role. *Journal of Philosophy* 74: 379–402.
Fodor, J. (1975) *The Language of Thought*. Cambridge, MA: Harvard University Press.
Harman, G. (1971) Three Levels of Meaning. In D. Steinberg and I. Jakobovits (eds), *Semantics*, Cambridge: Cambridge University Press: 66–75.
Harman, G. (1974) Meaning and Semantics. In M. Munitz and P. Unger (eds), *Semantics and Philosophy*, New York: New York University Press, pp. 1–16.
Katz, J., and P. Postal (1964) *An Integrated Theory of Linguistic Description*. Cambridge, MA: MIT Press.
Lewis, D. (1974) 'General semantics.' *Synthese* 27: 18–67.
Putnam, H. (1975) The Meaning of 'Meaning.' In K. Gunderson (ed.), *Minnesota Studies in the Philosophy of Science*, Vol. 7, Minneapolis: University of Minnesota Press, pp. 131–193.

2

Three Trivial Truth Theories

Ernest Lepore and Barry Loewer

According to Tarski, a theory of truth for a language L is a theory which logically implies, for each sentence S of L, a sentence of the form:

S is true in L iff *p*,

where ⌈S⌉ is replaced by a canonical description of a sentence in L and ⌈*p*⌉ is replaced by that sentence if L is contained in the metalanguage, or by a translation of S if L is not so contained (Tarski, 1956). Tarski constructed consistent and finitely axiomatized theories of truth for various formal languages and showed how to define 'is true in L' explicitly within these theories. We all agree that Tarski's theories of truth have enormous philosophical significance, but there is much less agreement on precisely what that significance is.[1]

The most eloquent and insistent champion of the philosophical importance of theories of truth is certainly Donald Davidson.[2] He has argued that truth theories are important because they can be employed as the core of a theory of interpretation of speakers of a language L; that is, anyone who knew the theory would be capable of specifying, for any indicative sentence in L, the content—or, to speak more intuitively, the significance—it would be taken to have by anyone who understands L (RI, p. 313). He has also claimed that finitely axiomatized truth theories for any natural language L will, as a matter of course, exhibit the recursive structure of L, and thereby it will both show how the meanings of complex expressions are composed of the meanings of their constituents and provide an account of the logical form of the sentences of L.[3] These claims, if true, would certainly establish the importance of truth theories. Here we want

---

[1] In work aimed mainly at philosophers, Tarski relates classical philosophical issues to his work in semantics. See Tarski, 1944, and compare also Tarski, 1969.

[2] Davidson has presented his views on these subjects in many places; see for instance his 1973a, 1973b, 1974, 1976, 1977. In the subsequent discussion four of these works will be referred to as IDCT (= 1973a); RI (= 1973b); BBM (= 1974); and RWF (= 1977).

[3] Davidson has made some rather strong claims on this point: "theories of absolute truth necessarily provide an analysis of structure relevant to truth and inference. Such theories yield a non-trivial answer to the question 'what is to count as the logical form of a sentence'" (IDCT, p 71.; cf. also BBM, p. 319; RI, p. 314). LePore (1982) argues that Davidson offers no good reasons for his claims about the relationship between a truth theory and a theory of logical form for a language. LePore attempts to provide one by realizing a connection in Davidson's views between truth and inference.

to examine a doubt about these claims which has been voiced by Belnap, Kripke, and Harman, among others.[4] These objections maintain that since constructing truth theories is all too easy, they cannot have the philosophical importance Davidson claims for them. Easy-to-construct truth theories appear to shed no light on recursive structure or on meaning. Consider [I]–[III]:

[I] (p) ('p' is true in L iff p)
[II] $(\varPi\ p)$ ('p' is true in L iff p)
[III] The theory whose axioms are the instances of "'p" is true in L iff p'.

In [I] the quantifier is objectual, while in [II] it is substitutional. Assuming that sense can be made of these theories, they are trivial in that they hold for any language L, irrespective of its structure. It is quite clear that none of these theories says anything about the logical structure of L, and it is not easy to see how they can make any contribution to a theory of interpretation for L.

In this chapter we will establish that a truth theory for a language L is the core of a theory of interpretation for L only in virtue of satisfying two constraints (BBM, p. 318; RF, p. 34):

(a) It entails, for each (indicative) sentence S of L, a true sentence of the form:

(T) S is true in L iff p

(b) It is an empirical theory, which—together with accounts of what speakers of L believe and desire and together with theories of rational belief change and action—successfully describes the linguistic (and other) behavior of the speakers of L.

We will then argue that, because the second of these requirements is neglected in the criticisms implicit in Kripke's, Belnap's, Harman's, and others' remarks, theories [I]–[III] cannot be used as the core of a theory of interpretation.

Before we take up these theories, we want to discuss a remark made by Tarski and repeated by many others, concerning

'p' is true in L iff p.

Tarski pointed out that, as quotes are usually understood, this sentence says that the sixteenth letter of the alphabet is true if and only if p, and that this is nonsense (1965, p. 160). This is, of course, correct but we can make sense of the scheme by employing quotes in a slightly non-standard way. There is no reason why we must interpret the quotation of 'p' as a name of the letter 'p.' We can instead stipulate that an expression consisting of a variable within quotes is not a name of anything, though the expression

---

[4] See for instance Belnap and Grover, 1973; Kripke, 1976; or Harman, 1974.

which results from replacing the variable with a term not containing free variables is a name. We return now to theories [I]–[III].

We are not alone in thinking something is wrong with these theories. Mark Platts, in a recent book which is a development and defense of the Davidsonian program, discusses each of these theories and argues, contra Kripke and others, that they are either incoherent or not really truth theories (Platts, 1979). If they were in fact incoherent or failed to be truth theories, then, of course, they would pose no problem for Davidson's claims that truth theories illuminate meaning. However, we will argue, against Platts, that [I]–[III] are truth theories. Platts says about [I] that,

> if we can overcome this first worry—say, by making sense of the idea that a used sentence names a state of affairs or perhaps a truth-value—we shall then encounter the problem of handling the mentioned occurrence of '*p*'. If the formula is to be well-formed, there must be one kind of object that both occurrences of '*p*' in the generalized bound formula can be treated as naming. Given the (supposed) way of handling the first worry, this would require that quotation marks be treated as expressing functions from states of affairs (or truth values) to sentences, the items of which truth is predicated; that is, they must be seen as operating upon an expression that names a state of affairs (or a truth value) to produce an expression that names a sentence. But this seems of very doubtful coherence. (Platts, 1979, p. 14)

It is not clear that this suggestion is incoherent. Let's suppose that the quantifier in [I] ranges over propositions. Sentences are construed as names of propositions, and 'if and only if' names a function which maps a pair of propositions onto a proposition which is true just in case the members of the pair agree in truth value;[5] single quotes name a function which takes a proposition as argument and yields a sentence which names that proposition as its value. The trouble with this suggestion is that, as propositions are usually construed, most languages contain many sentences which express the same proposition. If this is so for language L, then not all instances of form (T) will be forthcoming from [I]. We can remedy this situation, albeit somewhat artificially, by construing propositions so that each sentence of L expresses a different proposition— say, by identifying propositions with syntactic derivations of sentences. It may be difficult to take this theory seriously, but so far as we can see it does satisfy (T). We suppose someone might say that, if we are forced to introduce propositions or the like to make sense of [I], all the worse for it. But in our view what's wrong with [I] is deeper and more interesting than that. We will return to this point later.

Next we consider a substitutional interpretation of [II]. We suppose, of course, that the substitution class for '(Πx)' is the class of sentences of L. But this means that L must be contained within ML, the metalanguage of the theory. If we did not make this assumption, then either the sentences named on the left-hand side of instances of [II]

---

[5] In this chapter we will allow single quotes usually to name expressions, but sometimes they will be variously interpreted to make sense of the three trivial truth theories.

would not belong to L or the sentences on the right-hand side would not belong to ML. We can relax this a bit by appealing to translation:

(S) (S is true in L iff S is a sentence of L and (σ p) (Translation (S) = 'p' and p)),

where the first quantifier is objectual and the second is substitutional. Notoriously, substitutional quantification has its attackers and defenders. Attackers have claimed that substitutional quantification either is incoherent or is just objectual quantification in disguise. Defenders, most notably, Belnap, Grover and Kripke, have shown how to provide rigorous semantics for substitutional quantification. Here is what Platts says about [II]:

The interpretation of the universal quantification $\lceil (\Pi$ x) Fx$\rceil$ is this: all names, when concatenated with the predicate 'F' produce a *true* sentence. The problem is clear: substitutional quantification is defined in terms of truth, and so cannot itself be used to define truth. (Platts, 1979, p. 15)

But this is wrong. Substitional quantification is no more 'defined' in terms of truth than objectual quantification is. A truth definition for a language containing substitutional quantification does include a clause like:

For all terms t in the substitution class 'At/x' is true.

But this defines substitutional quantification in terms of 'true' no more than the objectual quantification clause

s satisfies $\lceil (x)A \rceil$ iff every sequence s', $S'_x = s'$, s' satisfies $\lceil A \rceil$

defines objectual quantification in terms of 'satisfaction'. In both cases the clauses are part of a definition of 'true-in-L', and not of a definition of quantification. It is the case that a truth theory for a language containing substitutional quantification contains among its basic clauses a characterization of truth for the language minus the substitution quantifier. But this doesn't show that substitutional quantification is defined in terms of truth either. It shows only that, to give a truth definition for a language containing substitutional quantification, truth must be characterized for sentences not containing the substitutional quantifier.

Even if [II] were defective as a *definition* of truth, because of some circularity, it still satisfies (T).[6] Davidson's position apparently is that truth theories do all the wonderful things mentioned at the beginning of this chapter but being a non-circular definition of truth was not one of these wonderful things. So [II] is still problematic vis-à-vis Davidson's claim since it is a truth theory, but it seems completely uninteresting as a theory of meaning. If it is to be rejected by Davidson, it must be for reasons other than those urged above.

---

[6] Davidson himself seems confused about this point. In IDCT he says: 'theories of truth based on the substitutional interpretation of quantification do not in general yield the T-sentences demanded by convention T' (pp. 79–80).

What about [III]? Platts says:

What of the residual possibility of treating ' "*p*" is true if and only if *p*' as a schema? Again this can be of no help in the enterprise of defining truth, for the appropriate notion of schema cannot be explained except in terms of truth. The claim would have to be that the result of replacing '*p*' in both its occurrences by one and the same sentence *p* is a true sentence of English. (Platts, 1979, p. 15)

It is difficult to determine what Platts has in mind. Is he claiming that the notion of an axiom schema cannot be explained except in terms of truth? Does this mean that the presentation of Peano's axioms, which employs an axiom schema, presupposes a notion of truth? Which notion? Truth in arithmetic? Of course, Peano (and other mathematicians) believe that the axioms are true in arithmetic; but the axioms do not presuppose the notion of truth. Platts seems to claim that, to explain the way an axiom schema is used to describe axioms, one must make use of the notion of truth. But this is not so. The explanation is that the axioms of a truth theory are each of the instances of ' "ρ" is true if and only if ρ'. The expression 'true' occurs nowhere in the explanation (though it occurs, of course, in the schema).

It is worth noting that Davidson gives passing attention to [III]. About it he says:

Such a theory would yield no insight into the structure of the language and would thus provide no hint of an answer to the question of how the meaning of a sentence depends on its composition. We could block this particular aberration by stipulating that the non-logical axioms be finite in number; in what follows I shall assume that this restriction is in force, though it may be that other ways exist of ensuring that a theory of truth has the properties we want. (Davidson, 1975, p. 19; compare BBM, p. 348)

So Davidson apparently uses brute force to reject [III], and perhaps he would reject theories [I] and [II] by brute force as well. But it would be interesting to see if there are reasons for rejecting these theories which flow naturally from Davidson's reasons for holding that truth theories can be used to interpret sentences of L. To do this, we will need to sketch Davidson's conception of interpretation.

According to Davidson, a theory of interpretation for a language L is a description of what an interpretor of L might know in as much as he understands L. But what does Davidson have in mind by such a theory? The project of interpretation is to construct a theory such that, given a suitable non-interpreting description of any possible utterance in L, the theory would enable anyone who knew it to assign correct interpretative descriptions to this utterance; that is, such knowledge would enable an interpreter to say, on a given occasion, what another's words from L meant.

In addition, a statement of the theory should not presuppose in any way an understanding of language L. This requirement is quite important, since the theory will fail to be empirical, in other words satisfy requirement (b), if it presupposes an understanding of L. For example, if the theory of interpretation of L used the concept of meaning-in-L unanalysed, it would be quite useless. This is why Davidson rejects Tarski's condition of adequacy, Convention T, which states that an adequate theory of truth for a language L is one which entails, for every (indicative) sentence of L, sentences of the form:

S is true in L if p,

where p is a *translation* of S (RI, pp. 316, 321). The problem with this requirement is that it presupposes translation rather than providing an account of when p translates S. Davidson's solution is to drop the requirement that p should translate S and to add the requirement that any theory which satisfies constraint (a)—which requires it to be empirical (that is, to satisfy constraint (b))—must be supported by certain evidence. Davidson is not crystal-clear concerning how a theory which entails 'S is true if and only if ρ' for each sentence S of L is supposed to account for the data—or even which data is supposed to be accounted for; but we assume he has something of the following sort in mind.

Esa, a German speaker, looks out the window and sees that it is snowing. He comes to believe that it is snowing and his belief is warranted, in ordinary circumstances, because he sees it is snowing. Blondie, our Interpreter, has her back turned to the window, but she, too, comes to share Esa's belief as a result of hearing Esa utter to her in German 'Es schneit'. What Blondie heard was some sounds, and this alone does not warrant her belief about snow. This scenario points to an important empirical condition on the interpretive project as Davidson conceives it: whatever set of propositions the Interpreter accepts, they must warrant (in part) certain specified beliefs acquired on the basis of hearing a speaker's utterance.

We might try to spell out Blondie's justification for her belief as follows:

(1) Esa utters and holds true the sentence 'Es schneit'.[7]

And suppose it is snowing.

One demand we make of justification is that proferred justifications make it reasonable that the conclusion obtains; but (1), by itself, does not do this. *At best*, on the basis of her belief that (1), it would be reasonable for Blondie to believe that 'Es schneit' is true, assuming she believes Esa to be reliable. One way we might try to strengthen the bridge between Esa's words and an interpreter's mind is to ascribe knowledge to Blondie of the conditions under which Esa's utterance is true:

---

[7] We use the expression 'utters' in the present context (in contrast to 'says that') in such a way that a speaker may utter (on a particular occasion) some words without his or our knowing what these words mean. On 'holds true,' Davidson says: 'a good place to begin is with the attitude of holding a sentence true, of accepting it as true. [. . .] It is an attitude an interpreter may plausibly be taken to be able to identify before he can interpret, since he may know that a person intends to express a truth in uttering a sentence without having any idea what truth. Not that sincere assertion is the only reason to suppose that a person holds a sentence to be true. Lies, commands, stories, irony, if they are detected as attitudes, can reveal whether a speaker holds his sentences to be true. There is no reason to rule out other attitudes towards sentences, such as wishing true, wanting to make true, believing one is going to make true, and so on, but I am inclined to think that all evidence of this kind may be summed up in terms of holding sentences to be true' (RI: 322).

(2) 'Es schneit' is true in German when uttered to me iff it is snowing at the time of utterance.

Given (1)–(2) and the assumption that Esa is reliable, it is reasonable for Blondie, our interpreter, to believe that it is snowing.

As many have argued, as speakers of a language we have the potential to produce indefinitely many sentences. This fecundity prevents a general description of an interpreter's linguistic competence in Esa's language which might consist solely of a list of truth conditions for each sentence she potentially interprets:

Blondie knows that 'Es schneit' is true iff it is snowing.
Blondie knows that 'Es regent' is true iff it is raining.
Blondie knows that 'Es ist Montag' is true iff it is Monday.

· · ·

· · ·

· · ·

We simply could never complete this list. But, if we are to state explicitly what an interpreter might know so that it would enable her to interpret another's words, we must put it in finite form. In this respect interpretation is no different from any other subject matter where we want not only a beginning, but an end as well. One strategy that suggests itself is to construct a theory which has, as consequences, sentences using the words 'is true if and only if' as a link between the description of a sentence and that sentence. An interpreter's knowledge consists of what is expressed by the theory; in particular, she knows the truth conditions for each sentence of the language. Such a theory enables us to characterize an infinite competence by finite means. Davidson intends truth theories to do this job. In proving, with the help of a truth theory, sentences of the form (T), we make use of finitely many axioms, which, together with familiar rules of inference, licence a series of substitutions, engineering a step-by-step shift, from words and structures mentioned on the left-hand side of the material bi-conditional to words and structures used on the right-hand side. In so doing, we derive, for each sentence of the language, a statement of the conditions under which it is true.

In addition, Davidson envisions a theory of truth for L qua theory of interpretation which is empirical and testable. It is empirical since the hypothesis that the T-consequences of the truth theory correctly characterize the linguistic competence of speakers of the language under investigation plays a role in interpreting their linguistic behavior (and other behavior, including their propositional attitudes). The theory is tested by seeing how good the description is. Our little story about Speaker Esa and her interpreter Blondie shows how the assumption that Blondie knows the truth conditions of 'Es schneit' in German explains why she is warranted in interpreting Esa's words, and therefore, under the right conditions, in believing that it is snowing when she hears this sentence uttered by Esa. However, there is something puzzling about the claim that these T-consequences are empirical.

Consider the famous

(A) 'Snow is white' is true in English iff snow is white.

Looked at in one way, (A) seems to be analytic, or true in virtue of its form alone. One
need not know the meaning of 'Snow is white' to recognize that (A) is true. On the
other hand, (A) seems to report a state of affairs which could easily have been
otherwise: namely, that the sequence of letters $\lceil S \rceil$ $\lceil N \rceil$ $\lceil O \rceil$ and so on is true if and
only if snow is white. The resolution is that (A) does report an empirical fact, but it does
so in a way that is bound to appear trivial to anyone who can interpret English. This is
not at all surprising, since interpreting English consists in part in knowing the fact
reported by (A). It is trivial because English is assumed to be contained in the
metalanguage: anyone who can interpret English will not only recognize (A) as true,
but also believe what it expresses. If this assumption is dropped, then (A) is not at all
trivial—it might even be false.

If the theory of truth is supposed to be part of an empirical theory of interpretation
for a language actually used by a group of people, it cannot be assumed that the
language in which the theory is expressed contains the language under study (IDCT,
pp. 83–84). Of course, the latter may be contained in the former, but this cannot be
assumed beforehand. This point can be expanded: if a truth theory is to be a theory of
interpretation for a language L, then understanding the articulation of the theory
should not itself assume an understanding of L. We will see that this is the sin of the
trivial truth theories. They satisfy requirement (a) all right, but they do so at the cost of
sacrificing their empiricality.

This is quite obvious with regard to [III], the specification of an infinity of
T-sentences by the axiom schema. One could, of course, know what sentences are
axioms of a truth theory for some language L without understanding L. The specification
of the axiom schema might be given in some other language. But one could not use
T-sentences derived from [III] to interpret sentences of L—that is, one could not know
what the T-sentences express—without already understanding L. Why? Because [III]
cannot be used to specify the T-sentences for L unless we articulate these sentences in
L. An example should help.

Let L be English. We want to show that, if we are to use [III] to articulate a truth
theory for L, then we must state [III] in English. Suppose we try to use German, so that
[III] becomes the theory whose axioms are the instances of:

'p' ist wahr, wenn und nur wenn p.

What can replace 'p'? If we use English sentences, we derive nonsense like the
following:

'It is snowing' ist wahr, wenn und nur wenn it is snowing.

If we use German sentences, then we succeed in specifying T-sentences for German and not for English; for

'Es schneit' ist wahr, whenn und nur wenn es schneit

does not translate the English sentence

'It is snowing' is true iff it is snowing,

but translates instead the English sentence

'Es schneit' is true iff it is snowing

—a sentence which purports to give truth conditions for the German sentence 'Es schneit' and not for the English sentence 'It is snowing'.

The upshot is that, if we employ specification [III] to describe the knowledge an interpreter might have, then we must state this theory in the language we are investigating. If we want to interpret German speakers, we must state [III] in German, and so on. This is the reason why [III] is useless as a specification of a theory of interpretation for L. What we want, supposing we are German speakers, is a German specification of T-sentences in which the left-hand side contains the name of a sentence of the language we are investigating, while the right-hand side contains a German sentence. What a French speaker would want is a French specification of T-sentences in which the left-hand side contains the name of a sentence of the language under investigation, and the right-hand side contains a French sentence, and so on. But the trivial [III] does not succeed in doing this.

The situation is similar with respect to theory [II]. Theory [II] satisfies (T) only on the assumption that L is contained in the language in which [II] is expressed. The reason is, obviously, that we must substitute the same sentence for the occurrence of 'p' both in and out of quotes, if we are to obtain an instance of (T) for [II].

Analagous remarks apply to the objectual version [I]. Let's suppose that the quote function assigns to a propsition a sentence of English. Then, if [I] is formulated in English, the theory is not empirical, since the object language is contained in the metalanguage. But even if [I] is formulated in another language, say, German, it can be understood only by someone who already understands English (if [I] is to be a theory for English). The reason for this is that, to understand instances of [I], we will need to know which English sentence is 'p', for propositions p. For example, let (B) be an instance of [I]:

(B) 'Es schneit' ist wahr, wenn und nur wenn es schneit,

where 'Es schneit' has as its value an English sentence (see footnote 5). In order to understand (B), we will need to know which English sentence is assigned to the proposition expressed by the German sentence 'Es schneit'. But, to know that this sentence is, say, 'It is snowing', we need to know that 'It is snowing' and 'Es schneit' name the same proposition. This is tantamount to assuming that we need to know a translation from German into English. If we know German, then this amounts to supposing that we understand English.

We cannot test the hypotheses that [I], [II], and [III] respectively characterize correctly the knowledge that various interpreters of L might have, since the assumption that they do so is embedded in their being meaningful. If our account is correct, then the trouble with the three trivial truth theories is not that they fail to satisfy Tarski's Convention T, but that they presuppose the truth of what they express. This is what makes them trivial and non-testable. In this respect they are like the utterance 'I exist': a boring triviality, since it presupposes what it expresses, though what it expresses is contingent and—for the utterer at least—an important fact.

We want to conclude with a few remarks concerning, first, the relationship between truth theories and translation and, next, the relationship between recursion and the principle that the meaning of the whole sentence is comprised of the meanings of its constituents.

We observed earlier that we can formulate a truth theory for L as follows:

(S) (Π p) ('p' = Translation (S) $\supset$ S is true in L iff p).

If we assume that each S in L has a truth-preserving translation in the metalanguage, then this theory satisfies Tarski's Convention T. This theory is not at all trivial, but it may appear that its non-triviality is due entirely to the translation function. It has been claimed that it is the workings of this translation function that will show the meanings of the sentences of L to depend on structure. The move from 'p' to p is quite trivial. This has suggested to Harman and Fodor that Davidson's insistence that a theory of meaning for L must satisfy Tarski's Convention T is unwarranted.[8]

A translation from L into ML will serve to explain the meanings of the sentences of L as well as a truth theory for L stated in ML would. But this is to forget that the aim of a theory of interpretation for L is to characterize the knowledge interpreters of L might have. And, whatever it is that they might know, which enables them to understand L, this knowledge is clearly not how to translate L into ML. As we argued earlier, what they might know, or part of what they might know, is the truth conditions of the sentences of L. The move from a translation of S into ML to a truth condition appears trivial to us, the speakers of ML, precisely because our understanding of ML consists in part in knowing the truth conditions of the sentences of L. The move from a translation of S into ML to a truth condition appears trivial to us, the speakers of ML, precisely because our understanding of ML consists in part in knowing the truth conditions of L.

Some readers might think that there is a rather easy and straightforward way to reject theories [I]–[III]. After all, since these theories are not recursive, they can shed no light on the structure of language L. But then how can they show how the meanings of

<hr />

[8]   See Fodor, 1975 and Harman, 1971 and 1974. A number of other authors have agreed with Harman and Fodor in seeing little difference, if any, between specifying meaning by translation and specifying meaning in terms of truth conditions—for example Field, 1977 and Putnam, 1975); See Chapter 1 for more discussion.

complex expressions are composed of the meanings of their constituents, and how can
they provide an account of the logical form of the sentences of L? Several authors have,
in fact, interpreted Davidson as requiring that an adequate theory of interpretation for a
language L must make use of recursion clauses (for example Tennant, 1977; Haak,
1978). Once such a requirement is introduced, it becomes possible that the sentences of
L will be articulated into singular terms, quantifiers, predicates, connectives, as well as
into the linking expressions with entities in the characterization of satisfaction (IDCT,
p. 84). As plausible as this rejection of theories [I]–[III] may seem, it is not one open to
Davidson.

Davidson does not come armed with a picture of how language is fragmented,
and therefore of how his solution for (T) is to go (IDCT, p. 79; RWR, p. 84). He
comes armed with intuitions about what it is to have an account of interpretation.
His own solutions involve utilizing theories which in fact include recursion clauses,
but this is a convenience and not a requirement. If there were a way of satisfying
constraints (a) and (b) without employing recursion, that would be perfectly
acceptable.[9]

It may turn out, as Davidson has intimated in several places that he believes (IDCT,
p. 81), that there is no way to satisfy Convention T without shedding light on the
logical and semantic structure of the language under investigation through the recursive structure of the theory. But this is something we would need to demonstrate, or
come reasonably to believe—at least inductively, through the repeated failure of truth
theories like [I], [II], and [III] to satisfy requirements (a) and (b). It is not something we
can assume beforehand.

In this chapter we have defended Davidson's position that a sound theory of
interpretation for L should satisfy (T) against charges of triviality. The moral we
drew from this exercise is that, although Davidson intends theories which satisfy (T)
to do a particular job, nothing he says implies that every such theory must do this job.
We have seen that certain trivial theories are not testable and that the difference
between truth theories and translation theories are not trivial. We suspect that the
illusion of triviality is engendered by our taking for granted our own language and
our own understanding of it. We have a tendency to take our own existence for
granted, appreciating it only once it is threatened. Perhaps the threats to truth
theories posed by the charges of triviality we have canvassed will lead us to appreciate
these truth theories.

---

[9] 'Convention T, in [the] skeletal form I have given it, makes no mention of extensionality, truth
functionality, or first order logic. It invites us to use whatever devices we can contrive appropriately to
bridge the gap between sentence mentioned and sentence used; restrictions on ontology, ideology, or
inferential power find favor, from the present point of view, only if they result from adopting Convention
T as a touchstone. What I want to defend is the convention as a criterion for theories, not any particular
theories that have been shown to satisfy the convention in particular cases, or the resources to which they
have been limited' (IDCT, 79).

Acknowledgments

An earlier draft of this chapter was presented at the Western American Philosophical Association meetings in Milwaukee, April 1981. We would like to thank Donald Davidson, Michael Root, and the editors of the *Canadian Journal of Philosophy* for their comments on that draft.

References

Belnap, N. D., Jr., and D. L. Grover (1973) Quantifying in and out of Quotes. In H. Leblanc (ed.), *Truth, Syntax and Modality*, Amsterdam: North-Holland Publishing Co., pp. 17–47.

Davidson, D. (1973a) In Defense of Convention T. In H. Leblanc (ed.), *Truth, Syntax and Modality*, Amsterdam: North-Holland Publishing Co., pp. 76–86, reprinted in Davidson (2001)[= IDCT].

Davidson, D. (1973b) Radical Interpretation. *Dialectica* 27: 314–328, reprinted in Davidson (2001) [= RI].

Davidson, D. (1974) Belief and the Basis of Meaning. *Synthese* 27: 309–323, reprinted in Davidson (2001) [= BBM].

D. Davidson (1975) Semantics for Natural Language. In D. Davidson and G. Harman (eds), *The Logic of Grammar*, New York: Dickenson Publishing Co. Inc., pp. 55–64.

Davidson, D. (1977) Reality without Reference. *Dialectica* 31: 247–258, reprinted in Davidson, D. (1976) Reply to Foster. In G. Evans and J. McDowell (eds), *Truth and Meaning: Essays in Semantics*, Oxford: Clarendon Press, pp. 33–41, reprinted in Davidson (2001).

Davidson (2001) [= RWR].

Davidson, D. (2001). *Inquiries into Truth and Interpretation*, 2nd edition, Oxford: Oxford University Press.

Field, H. (1977) Logic, Meaning and Conceptual Role. *Journal of Philosophy* 74: 379–402.

Fodor, J. (1975) *The Language of Thought*. New York: Thomas Y. Crowell Company.

Haack, R. J. (1978) Davidson on Learnable Languages. *Mind* 87: 230–249.

Harman, G. (1971) Three Levels of Meaning. In D. Steinberg and I. Jakobovits (eds), *Semantics*, Cambridge: Cambridge University Press, pp. 66–75.

Harman, G. (1974) Meaning and Semantics. In M. Munitz and P. Unger (eds), *Semantics and Philosophy*, New York: New York University Press, pp. 1–16.

Kripke, S. (1976) Is There a Problem about Substitutional Quantification? In J. McDowell and G. Evans (eds), *Truth and Meaning*, Oxford: Clarendon Press, pp. 325–419.

Putnam, H. (1975) The Meaning of "Meaning." In K. Gunderson (ed.), *Minnesota Studies in the Philosophy of Science*, Vol. 7, Minneapolis: University of Minnesota Press, pp. 131–193.

Lepore, E. (1982) Truth and Inference. *Erkenntnis* 18: 379–395.

Platts, Mark (1979) *Ways of Meaning*. London: Routledge and Kegan Paul.

Tarski, A. (1944) The Semantic Conception of Truth. *Philosophy and Phenomenological Research* 4: 341–375.

Tarski, A. (1965) The Concept of Truth in Formalized Languages. In A. Tarski, *Logic, Semantics, and Metamathematics*, Oxford: Clarendon Press, pp. 152–278.

Tarski, A. (1969) Truth and Proof. *Scientific American* 220: 63–77.

Tennant, N. (1977) Truth, Meaning and Decidability. *Mind* 86: 368–369.

3

What Model-Theoretic Semantics Cannot Do

Ernest Lepore

It has frequently been argued that structural semantic theories (for instance Katz, 1966, 1972; Katz and Fodor, 1963; Katz and Postal, 1964; Jackendoff, 1972) and some versions of generative semantics (Bach, 1968; Lakoff, 1972; McCawley, 1968) are deficient in an essential way. Cresswell (1978), Lewis (1972), Partee (1975), and Vermazen (1967) all argue (among others) that structural semantic theories (hereafter, SS) do not articulate relations between expressions and the world, that they do not provide an account of the conditions under which sentences are true, and that, therefore, these theories are not really semantics. In their place, many philosophers and linguists endorse model-theoretic semantics (hereafter, MTS) (see Cresswell, 1973; Lewis, 1972; Montague, 1974a, 1974b, 1974c; Partee, 1975). They do so because they believe that MTS compensates for what is deficient in SS. My aim in this discussion is to reconstruct the case against SS by demonstrating that the concept of truth is central to semantics and that a theory which issues in truth conditions for sentences of a language L must be the heart of a semantic theory for L. But I will also argue that MTS theories by themselves, somewhat surprisingly, are inadequate in exactly the same way as SS theories. If I am correct, then the widespread view that MTS can provide either a theory of meaning or a theory of truth conditions for the sentences of a natural language is mistaken.

1. Structural semantic theories

SS theorists countenance properties and relations like synonymy, antinomy, meaningfulness, meaninglessness or semantic anomaly, redundancy, and ambiguity as a good initial conception of the range of semantics. They do so because, for them, a semantic theory for a language L is a theory of meaning for L, and they believe that properties and relations like these are central to our concept of meaning. Therefore, any theory which did not bear on all, or at least many, of these phenomena should be suspect as a semantic theory (Fodor, 1977; Katz, 1972).

Theories in the SS vein proceed by translating or mapping natural language expressions into (sequences or sets of) expressions of another language. There is no uniformity among SS theorists about the nature of this other language, or about how these translations or mappings are to be effected. (For further discussions of these theories and their differences, see Fodor, 1977.) For our purposes, we beg no questions by restricting attention to Katz's suggestion, whereby the language translated or mapped into is 'Semantic Markerese' (Katz, 1966, 1972; Katz and Fodor, 1963; Katz and Postal, 1964). The culmination of the various translation rules and other apparatus within Katz's theory results in theorems like:

(A) 'Barbara sekoilee' in Finnish translates into the language of Semantic Markerese as S.

Translations of this kind are constrained, and this is the reason for bringing them in along with the semantic markers in the first place, such that synonymous expressions of some language L translate into the same (sequence or set of) expressions of Semantic Markerese, ambiguous expressions of L translate into different expressions of Semantic Markerese, anomalous expressions of L translate into no expressions of Semantic Markerese at all, and so on. Facts about synonymy, ambiguity, anomaly, and other semantic properties and relations are accounted for through these representations, translations, and semantic constraints (or definitions).

Some critics of Katz's theory charge that the phenomena he concerns himself with represent only a sample of the full range of facts semantics must ultimately deal with, and they argue that SS cannot in principle accommodate this full range. In particular, some argue that it is the construction of truth conditions that should count as the central concern of semantics, not these other properties and relations, and that SS theories cannot provide truth conditions (Cresswell, 1978; Davidson, 1967, 1973; Hintikka, 1969; Lewis, 1972; Montague, 1974a, 1974b; Partee, 1975; Stalnaker, 1972). This raises two questions: why can't SS theories provide truth conditions, and why should they? This second question is especially significant inasmuch as SS theorists have expressed bewilderment in the face of the criticism that their theories do not specify truth conditions.

Katz, for one, agrees that his semantic theory leaves out the notion of truth, and therefore does not specify truth conditions. But, he goes on to say, 'the subject matter to which "truth" is central is not one that my semantic theory is or was ever intended to be about' (1972, p. 182). If 'semantics' is construed as having to do with meaning, then truth 'is not central to semantics and so there can be no claim that my theory has left out something central' (ibid.). As Katz sees it, the criticism that his theory does not specify truth conditions trades off an ambiguity of the term 'semantics.' For SS theorists the goal of a semantic theory is to construct a theory of meaning, whereas for their critics the goal of a semantic theory is 'to study relations between objects of one sort or another and the expressions of a language that speak about them' (ibid., p. 183), relations which ultimately get spelled out through truth conditions for sentences of this language. Therefore, if the criticism that SS theories do not specify truth conditions is to carry any force with Katz, we must show that a semantic theory for a natural language which has as its goal the construction of a theory of meaning for this language must, in order to achieve this goal, specify truth conditions for sentences of this language. This would have the effect of collapsing Katz's two senses of 'semantics' into one. We take a step in this direction by asking: what should we expect from a semantic theory as a theory of meaning?

Traditional wisdom about meaning is that it is in virtue of knowing what a sentence means that we (in part) understand it. For example, it is only in virtue of knowing what 'Barbara sekoilee' means that I am warranted in believing that an assertive utterance of these words is an assertion that Barbara is confused. If I further know that these words are true on the occasion of utterance, then knowing their meaning warrants (in part) my believing that Barbara is confused as well.[1]

This illustration brings out nicely the two-sidedness of the concept of meaning. On the one hand, meaning is connected with a host of extensional concepts: satisfaction, denotation, truth, and so on. This is reflected in the principle implicit in our example: if a sentence S is true, and if S means that p, then p. On the other hand, meaning is connected with a host of intensional concepts: indirect quotation, assertibility, and so on. This is reflected in the connection we just saw between meaning and indirect quotation: if someone assertively utters a sentence S, and S means that p, then this person says that p. Given the meaning of a sentence, this duality permits us to move in either of two directions. We can exploit the relationship between meaning and truth to infer something about the world beyond the speaker, or we can exploit the relationship between meaning and certain intensional notions to infer something about the speaker himself, what he asserted, queried, commanded, and so on. With this wisdom in mind, we expect a semantic theory—as a theory of understanding (or as a theory of semantic competence) for a language L—at least to specify the meanings of sentences of L.[2] SS theorists agree about this, but they have erred in assuming that any semantic theory which accounts for semantic properties and relations like ambiguity and synonymy in the manner suggested above will as a matter of course also specify meanings for sentences in an appropriate way.

Any SS theory which issues in (A) certainly entails:

(B) 'Barbara sekoilee' means the same as S.

Therefore any SS theory which issues in (A) can be said to specify the meaning of 'Barbara sekoilee', but not in an appropriate way, since (B) alone will not warrant its knower in believing that an assertive utterance of the sentence named on the left is an

---

[1] For arguments supporting these claims, see Lepore, 1982; Lepore and Loewer, 1981.

[2] One large issue that I will not address is the issue of whether the linguist's conception of a 'competence theory' can be satisfied either by a structural semantics or by a model-theoretic semantics. The problem is how to characterize what's in the 'speaker's head'. For MTS, the issue is the status of the model theory; for SS, the issue is the status of the metalanguage in which the mappings are given.

assertion that Barbara is confused. Nor would its knower be warranted in believing that Barbara is confused if he further knew that this sentence is true. Why not?

In the overall picture of SS there are three languages: the natural language, the language of Semantic Markers, and the translating language (which may be Semantic Markerese, the natural language, or some other language). Translation proceeds by correlating the first two of these using the third. But it is possible to understand (A) or (B) knowing only the translating language (in this case English), and not the other two. Put somewhat differently, we can know that a sentence translates, or means the same as, another without knowing what either one means. We can perhaps know that (A) or (B) is the case on the basis of what Katz tells us, without knowing what either 'Barbara sekoilee' or the Semantic Markerese sentence S means.

Of course, if someone understands Semantic Markerese, then she can no doubt use (B) to interpret the Finnish sentence. But this is because she brings to bear two things she knows that (B) does not state: the fact that Semantic Markerese is a language she understands; and her particular knowledge of how to interpret S. This latter knowledge is doing most of the work here—not the SS theory. And it is this knowledge that we want an adequate semantic theory of meaning to characterize.

Nothing we have said so far, however, establishes either that a semantic theory as a theory of meaning should be concerned with truth conditions or, for that matter, that SS theories cannot be used to assign truth conditions. In fact, there is some prima facie evidence to think that SS theories can be used to assign truth conditions to sentences. For, surely, if (A) holds, then it follows that (C):

(C) 'Barbara sekoilee' is true in Finnish if S is true in Semantic Markerese.

Since S's being true in semantic Markerese is one condition under which 'Barbara sekoilee' is true in Finnish, shouldn't (C) count as providing truth conditions for the Finnish sentence? It would seem that, if there is a deficiency in SS theories with respect to truth conditions, then something must be wrong with the kind of specification of truth conditions (C) provides. To see that this is so, we must ask why a semantic theory as a theory of meaning should be concerned with truth conditions in the first place.

We said before that someone who knows the meaning of 'Barbara sekoilee' would, presumably, be warranted in believing that Barbara is confused if he further knew these words to be true. But this is exactly what we would expect someone to be licensed to believe if he knew the conditions under which the sentence is true. The sentence is true if and only if Barbara is confused. That is to say, at least for a straightforward declarative sentence, in specifying the conditions that have to hold for it to be true, we are in effect characterizing a central aspect of its meaning.

Seen from another angle, suppose that someone knows the meaning of 'Barbara sekoilee' and knows all the relevant facts (or, not to be tendentious, knows everything in the *world* there is to know): then this person will know whether the sentence is true. How could this be, unless meaning determined truth value throughout the relevant possible states of affairs? And, if meaning does determine truth value in this way, then a theory of meaning for a language will have to specify truth conditions. (Indeed, many semanticists would go so far as to say that knowledge of truth conditions for a sentence *is* knowledge of its meaning; see for example Cresswell, 1978. They do so because they believe that knowledge of truth conditions warrants whatever the knowledge of meanings has traditionally been thought to warrant.)

If the meaning of a sentence includes as a part (or is identical to) truth conditions for that sentence, then any semantic theory for a language which purports to be a theory of meaning for that language must specify the truth conditions for each sentence of the language. From this it does not follow that *any* semantic theory which provides a complete specification of truth conditions for sentences of that language is adequate. An SS theory which issues in theorems like (A) may entail (C), but (C) alone does not warrant its knower to believe that Barbara is confused if he further knows that 'Barbara sekoilee' is true. This is because (C) does not specify truth conditions for 'Barbara sekoilee' in an appropriate way.

What we have shown is that SS theories, though they may account for some aspects of our concept of meaning, cannot account for them all. For whatever knowing the meaning of an expression includes, it does not involve simply translating the expression into a semi-formal language, or telling us which other sentence it means the same as, or telling us which other sentence it has the same truth conditions as. We turn now to MTS with an eye towards how it can compensate for what is deficient in SS, i.e. how is MTS able to provide an explicit characterization of what SS assumes and leaves unsaid?

2. Model theoretic semantics

Model theory has traditionally been used as a mathematical technique for investigating certain properties of formal systems such as consistency, completeness, the finite model property, and having a decision procedure. There is now a growing impression among linguists and philosophers that model theory can provide a theory of meaning for natural languages. This view has greatly come into its own in the last few decades largely because of the work of Kripke (1963a, 1963b), van Fraassen (1969), Hintikka (1969), Montague (1974a, 1974b, 1974c), Lewis (1972), and others. These authors have developed MTS for formal systems of many valued, sortal, free, tense, demonstrative, counterfactual, and modal logics. These results have encouraged many researchers to believe that MTS may be sufficiently powerful to provide a theory of meaning for substantial fragments of natural language. There are many competing approaches; each, however, seeks to characterize (or define) a relativized concept of truth (at a world, time, or whatever other index is deemed relevant). Here I will focus my discussion on Montague's grammar, in particular on the theory presented in Montague (1974b) (hereafter, PTQ). What I have to say about PTQ extends obviously to any MTS approach. Montague personally was not interested in a theory of understanding. My discussion, however, is directed, not at Montague, but instead at those semanticists who are interested in formal semantics as a theory of understanding and at those who have argued that MTS constitutes a real advance over SS. I focus on PTQ because it is familiar and because many semanticists who are interested in semantic competence employ the theory of PTQ or some variant of it (Potts, 1975).

In PTQ, Montague proposes a general theory of syntax and MTS. He treats a fragment of English which includes simple quantification and some intensional verbs. His theory involves three distinct phases. English expressions are assigned a syntactic analysis with respect to a categorial grammar. This syntax is translated into the syntax of a tensed intensional logic with various non-logical constants. Finally, the expressions of this intensional logic undergo a model-theoretic interpretation. This interpretation proceeds by linking linguistic entities with non-linguistic entities in two ways: method of extension and method of intension.

The extension of an expression from some language L is determined relative to an interpretation A of L and a world w and time t in A (i.e. relative to the model $<A <w, t>>$ of L). In short, it is the object the expression denotes in A at w and t. The intension of this expression is the meaning, sense or concept correlated with the expression. Instead of treating intensions as basic, as some kind of ideal abstract entity or mental representation, Montague defines the intension of an expression as a function: it is the function which, for every possible world w and time t (in A), picks out exactly those objects in A which make up the extension of this expression in A at w and t. We need not go into any great detail. Suffice it to say that the culmination of the various definitions, translations, rules and other apparatus within PTQ result in theorems like the following:

(E) 'Barbara sekoilee' is true in an interpretation A at a world w and a time t (in A) if and only if the extension picked out by the intension of 'Barbara' in A at w and t is a member of the extension picked out by the intension of 'sekoilee' in A at w and t.

(I) The intension of 'Barbara sekoilee' in an interpretation A is a complex function from the set of possible worlds and times in A onto the set of truth values, true and false, where this complex function is arrived at by composing the intension of 'Barbara' with the intension of 'sekoilee'. The intension of 'Barbara' is a function from possible worlds and times (in A) to individuals in A. Similarly, the intension of 'sekoilee' in A is a function from possible worlds and times (in A) to functions from individuals in A to truth values (alternatively, one can say, from possible worlds and times to classes of individuals). (Montague has different intensions for proper names, and he treats predicate intensions as arguments of proper name intensions in composing the two functions. Neither of these points, however, affects the present discussion. I have chosen these intensions of expository purposes.)

It is held by many MTS theorists that there are important advantages PTQ offers over its SS competitors (see Cresswell, 1978; Lewis, 1972; Montague, 1974a; Partee, 1975). It is distinguished from an SS approach because it links expressions of one language with expressions of another (that is, instead of stopping at phase two in PTQ), whereas PTQ links expressions to non-linguistic entities (the third phrase in PTQ). Our question is: why should these links signpost an advance over SS? What advantages accrue to PTQ in virtue of having consequences like (E) and (I), which do not accrue to SS theories?

The PTQ embodies some very special claims about the fundamental nature of semantic interpretation and about the way in which syntax and semantics systematically correlate. This correlation embodies the familiar Fregean principle of compositionality: stated crudely, this is the principle that the meaning of the whole is a function of the meanings of its parts. The correlation is realized in Montague's work by giving the syntax the form of a simultaneous recursive definition of the sets of well-formed expressions of each syntactic category of the language, recursively building up larger phrases and clauses from smaller ones, and associating with each syntactic formation rule a semantic interpretation rule that specifies the interpretation of the constitutent phrases. Most semanticists argue that an adequate theory of meaning must embody this kind of compositionality. Otherwise it would be impotent to account for the obvious and essential fact that we can understand hitherto unencountered sentences.[3]

Evaluation of a semantic theory for some language is not limited to whether or not the semantic theory embodies a principle of compositionality alone. It involves also, to some extent, evaluating how accurate a map the theory provides of the logical geography of the language, that is, of the logical consequences, truths, equivalences, and other logical properties and relations. After all, part of understanding a language involves knowing which sentences stand in logical relationships (like logical consequence) to others. Someone who did not know that, whenever a sentence of the form ⌜P and Q⌝ is true then sentence ⌜Q⌝ is true, cannot be said to understand English—or at least one important word in English, 'and.' One important benefit MTS offers is a way to define these important logical notions. (If we consider a subset K of the models determined by an interpretation for some language, we can define a sentence O as K-valid if it is true in each of these models in K. If K is the set of models (determined by any interpretation) in which all the logical words of the language (such as 'not', 'or', 'and') receive the extensions usually given by logicians to these words, then these will be the logically possible models for L. Then K validity is logical validity. In PTQ, Montague effects this restriction by a set of meaning postulates. Meaning postulates are ways of placing restrictions on the interpretation of expressions. We could have a notion of logical validity based on the subset of models in which all of the meaning postulates are true (Montague, 1974b, p. 263; 1974c, p. 236).

Giving an account of compositionality and one of logical consequence, therefore, are two central goals for PTQ. But—and this is a big 'but'—these two goals are also central for SS. Katz seeks to embody a Fregean compositionality in his theory. His dictionary assigns a meaning ('lexical reading') to each basic expression of the language. Projection rules in his theory can be regarded as semantic operations, where

---

[3] An exception to this compositionality principle is game-theoretic semantics. See Hintikka and Carlson, 1979; Saarinen, 1977; See also Kamp (1981).

there is a projection rule corresponding to each phrase structure rule. These projection rules combine recursively the readings for each node immediately dominating lower nodes (Katz, 1972; Katz and Postal, 1964). The more vital notion of entailment (and the family of notions definable in terms of it) Katz attempts to define not in terms of classes of models (nor in terms of inference in some formal deductive system), but rather in terms of containment of (parts of) one reading in another (Katz, 1972, 1975, 1977; Katz and Nagel, 1974).

None of the arguments in the critical literature shows that Katz's theory cannot in principle accommodate these two essential semantic features. And, what's more important—and this is very important—the translation argument, the argument MTS theorists themselves have preferred in criticism of SS, was certainly not proferred to show that SS cannot account for compositionality or logical consequence. This is so because, just as I can know that one sentence translates another without understanding either, I can know that the first entails the second without understanding either, without knowing what either means (though, perhaps, knowing that one sentence implies another is part of knowing what each sentence means). And, also, I can know how the parts of expressions combine to issue in the meanings of the larger expression without understanding this expression. If all we wanted from a semantic theory were to account for *these* two aspects of language, then no reason in principle has been proferred for preferring MTS over SS. Therefore, if SS is deficient in a way that MTS is not, then there must be another aspect of language that an adequate semantic theory must address, which MTS addresses and SS does not.

The deficiency critics of SS emphasize that SS fails to provide a connection between expressions and extralinguistic entities. Barbara Hall Partee, for example, writes:

Semantics [à la Montague and Thomason] has always been the study of the relations between expressions in a language and the non-linguistic subject matter that the expressions are about [...] No amount of [...] interlinguistic connections can serve to tie down the extralinguistic content of intensions. For that there must be some language-to-world ground. (Partee, 1980, pp. 69–70)

Quotes of this sort can be produced ad infinitum. According to these authors, PTQ represents an advance over SS because it *requires* realizing a connection between expressions and extralinguistic entities. Contrary to proponents of SS, MTS proponents intend to break out of the 'confines of language'. Earlier we saw the importance of having a semantic theory that offered an explicit and general way of accounting for the relationships that hold between any sentence and a situation, where the sentence's truth conveys information about that situation. In virtue of understanding the sentence 'Barbara sekoilee', I can come to have a belief about the non-linguistic world, namely the belief that Barbara is confused, upon hearing this utterance. A theory which never moves beyond mentioning language cannot accommodate this feature of language, for it is only by using language that we can talk about a non-linguistic world. But, still, we must ask whether PTO issues in theorems which engineer transitions from utterances to assertions. We have no reason beforehand to assume that any linguistic/non-linguistic link a theory forges will license ascriptions characteristic of language understanding. I will now argue that PTQ, like SS, does not provide enough to bridge the gap between utterance and assertion.

Suppose that Frank utters the words 'Barbara sekoilee', and all I know about Frank's language is that (E') and (I') hold:

(E') 'Barbara sekoilee' is true in Finnish iff whatever 'Barbara' picks out is one of the things 'sekoilee' is true of.

(I') The meaning of 'Barbara sekoilee' in Finnish is the proposition which results from taking the meaning of 'Barbara' as argument for the meaning of 'sekoilee'.

It would be quite remarkable if I were able to discern what Frank asserts when he utters 'Barbara sekoilee', provided that (E') and (I') constituted the whole of my knowledge about Frank's language. Knowing that (E') or that (I') at best warrants my believing that Frank asserted that something named 'Barbara' has the expression 'sekoilee' true of it. It would remain a mystery to me which thing it is, and exactly what is true of it.

Unless a sentence uttered is about language, reference to language must be eliminated entirely by any semantic theory which seeks to provide a theory of meaning. Whatever we come up with has got to take its knower from the perception of the sequences of sounds to their characteristics. Once you understand the motivation for this condition, you should see that PTQ, apart from whatever other riches it may yield, is as inadequate as SS. Both take a wrong direction from the start. We are never told straight out what the truth conditions or meanings of sentences are. Instead of an interpretation of names and predicates being fixed these are left open in PTQ. The notions of truth and denotation are defined relative to a given interpretation, which includes a given set of possible individuals, worlds, and times. (E) and (I) tell us how to derive the truth conditions of, and the proposition expressed by, the sentence 'Barbara sekoilee' only relative to an interpretation A and to a given world w and time t in A. In order to complete the disquotation and get at the actual truth conditions and propositions expressed by this sentence, we need to specify (single out) the actual interpretation and the actual world and time. In this regard it is illuminating to compare PTQ with Davidson's semantic theory.

Davidson's theory differs from PTQ in at least one important respect. Davidson (1967) argues that the kind of structure needed to account for language understanding is either identical with or closely related to the kind given by a definition of truth along the lines first expounded by Tarski (1956). Such a theory (by means of a set of axioms) entails, for every sentence in the language, a statement of the conditions under which it is true, expressed by bi-conditionals of the form 'S is true if and only if p'. (S here refers to the sentence whose truth conditions are being given, and p is that sentence itself. If the language in which these truth conditions are being stated does not include the language to which S belongs, then p will have to be a translation of S.) This condition of adequacy excludes MTS theories, because these theories move in a direction different from that proposed by Davidson's conditions. Since MTS theories substitute a relational concept for the single-place truth predicate, such theories cannot carry through the last step of the recursion of truth (or satisfaction), which is essential to the quotation lifting feature of the truth-conditional bi-conditionals (Davidson, 1973).

Put somewhat differently, from a relativized truth theory we cannot derive an absolute truth theory. Thus, suppose, for example, language L consists of one sentence, 'Barbara sekoilee'. An MTS theory for L along the lines of PTQ would issue in a theorem somewhat like:

(1) $(A)(p)$ ('Barbara sekoilee' is true in A at p iff the extension of 'Barbara' in A at p satisfies 'sekoilee' in A at p),

where A ranges over interpretations and p ranges over possible worlds. (We omit reference to times.) From a relativized truth theory for L like (1) we cannot derive an absolute truth theory for L like:

(2) 'Barbara sekoilee' is true iff Barbara is confused.

The weakest addition to (1) we could make in order to derive (2) from it is the following:

(3) $(\exists A)(\exists p)(x)$ ((the extension of 'Barbara' in A at p = Barbara) & (x satisfies 'sekoilee' in A at p iff x is confused) & ('Barbara sekoilee' is true in A at p iff 'Barbara sekoilee' is true))

The first two clauses in (3) essentially say that, in order to understand L, we must know, in addition to (1), the base clauses in an absolute truth theory for L, that is:

(4) The extension of 'Barbara' = Barbara,
(5) (x) (x satisfies 'sekoilee' iff x is confused).

The last two clauses in (3) essentially say that we must further know that interpretation A is the actual interpretation and some world p in A is the actual world. Knowing that A is the actual interpretation and p is the actual world allows us to infer that truth in A at p is absolute truth.

What can we conclude from the fact that (1) alone does not imply (2)? Hartry Field raises this question in other context (Field, 1973): he asks whether the fact that, for example, (6) contains a semantic term where (4) does not gives theories which employ base clauses like (4) an advantage over those which employ base clauses like (6):

(6) 'Barbara' denotes what it denotes.

After a long investigation, in which he finds no adequacy condition on an absolute truth theory which rules out (6) but not (4), he concludes that there is unlikely to be any philosophical purpose or interest to the claim that theories which employ clauses like (4) serve better than those which employ clauses like (6). Partee endorses Field's conclusion. She writes (1977, pp. 321–322):

As Hartry Field argued, a Tarskian truth definition has at its basis a listing of denotation conditions for the primitive terms in the form (7), and for the primitives we might just as well start with the form (8).

(7) 'Snow' denotes snow.

(8) 'Snow' denotes what it denotes.

Partee agrees with Field (1973) (and with Harman, 1974) that '[t]he real work of the truth definition, and similarly for a Montague-style possible world semantics, comes in the specifications of how the interpretations of the infinite set of sentences can be determined by a finite set of rules from the interpretations of the primitives' (1977, p. 322.). The idea here is that truth-conditional semantics illuminates meaning not by assigning truth conditions but through exhibiting the roles of logical words 'and', 'or', and the like in its recursive clauses. Partee sees Tarski-style truth theories and MTS style theories as being equally inadequate when it comes to specifying the meaning of lexical items.

One author, Richmond Thomason, taking an uncharacteristically Quinian line, goes further by arguing that we *cannot* reasonably expect semantic theories to tell us anything important about the meaning of lexical items. He writes:

The problems of a semantic theory should be distinguished from those of lexicography [. . .] A central goal of (semantics) is to explain how different kinds of meanings attach to different syntactic categories; another is to explain how the meanings of phrases depend on those of their components [. . .] But we should not expect a semantic theory to furnish an account of how any two expressions belonging to the same syntactic category differ in meaning. 'Walk' and 'run', for instance, and 'unicorn' and 'zebra' certainly do differ in meaning, and we require a dictionary of English to tell us how. But the making of a dictionary demands considerable knowledge of the word. (Thomason, 1974, pp. 48–49)

To ask the semanticist to give a specification of the meanings of words would be to ask too much, since it would require of him that he construct a world encyclopedia.

Each of these authors has gone wrong because he or she has failed to appreciate the differences between an absolute truth theory and MTS. First, this can be seen with regard to Partee's and Field's claim that there are no advantages in absolute truth theory with base clauses like (4) over a theory, for instance an MTS theory, with base clauses like (6). Simply note that, if these authors were right, then a Davidsonian truth theory would tell us no more about lexical semantics than Montague's theory. But we have shown that this is false. Adding (6) to PTQ will not license the kinds of reasoning we have been probing, reasoning we have argued is characteristic of language understanding. Adding (4) will. (6) does not eliminate reference to language in ways that (4) does.

What about Thomason's argument that lexical semantics is not part of semantics proper, since to distinguish the meanings of any two terms frequently requires more information than we can reasonably expect a semantic theory to provide us with? We cannot reply to him that a semantic theory which does not specify the meanings of the lexical items of a language L fails to specify the knowledge requisite for understanding L. Thomason's position apparently is that we cannot reasonably expect a semantic theory to specify all this knowledge. Semantics proper, according to him, is to specify the meanings of the connectives involved in inference. But Thomason is wrong here. Why should we think that the specification of the knowledge required for understanding lexical items in our language demands as much knowledge as Thomason thinks? Put somewhat differently, what do we expect to achieve by eliminating reference to language in the base clauses of PTQ? Again, we want a theory which will provide truth conditions (and/or meanings) of sentences of the language in such a way that someone who knows these truth conditions (or meanings) would be licensed to believe what the speaker asserted about the world. Our question is: what do we need to know about the difference (to borrow Thomason's own example) between the words 'run' and 'walk' to guarantee such competence? Presumably clauses like (9) and (10):

(9) (x) (x satisfies the predicate 'run' if x runs).
(10) (x) (x satisfies the predicate 'walk' iff x walks).

Someone who had this knowledge would be licensed to believe that Barbara runs when he hears the words 'Barbara runs' uttered by a reliable speaker, and he would be licensed to believe that Barbara walks when he hears the words 'Barbara walks'. This knowledge does not seem at all to require considerable non-linguistic knowledge. In fact, if anything, it seems to be paradigmatic of linguistic semantic knowledge.

What does all this add up to? One response is to say that our results are unsurprising since MTS is primarily valuable as a theory of logical consequence rather than as a theory of meaning. These two kinds of theory have different goals, which have induced salient differences in approach. A theory of logical consequence is concerned with the validity of forms of argument, represented by inference schemas. Therefore it must attend to a multiplicity of possible interpretations of a sentence schema: the notion it requires is that of truth under an interpretation. A theory of meaning, as we have portrayed it here, is concerned only with a single interpretation of a language, the correct or intended one: so its fundamental notion is that of meaning or truth—simpliciter.

However, despite their differences in goals, these two theories have been closely allied historically. Throughout the subsequent course of both subjects, theorists of meaning have borrowed from theorists of logical consequence many of the concepts devised by logicians, MTS being a primary example. The differences in goals between the two subjects raises the question: How far can the devices employed by logicians be made to serve the different purposes of semanticists of natural languages? The upshot of our investigation is that MTS can serve them no better than, and is consequently as deficient as, SS.

3. The proper form of a semantic theory

Many Montague grammarians and other proponents of MTS would agree with these last points but would argue that, in characterizing the collection of all interpretations of a language, we also do specify a particular one, the actual interpretation, and that this specification of the actual interpretation would also involve specifying the actual world, thus enabling us to characterize absolute truth. After all, Tarski himself describes MTS as the general theory of which the absolute theory is a special case (1956, p. 156). Montague showed no inclination to single out a unique interpretation of English, but he did note that 'not all interpretations of intensional logic, however, would be reasonable candidates for interpreting English' (1974b, p. 263). And in Montague (1974a) he says:

To be specific, a sentence would be considered true with respect to an analysis or a possible world i if it were true (in the sense given earlier) with respect to the actual model and i. This relativization to i would be eliminable in the same way once we were able to single out the actual world among all possible worlds. (p 209)

The question we have been pursuing here is what *form* a semantic theory should assume, what kinds of procedures for presenting meaning and truth conditions should a semantic theory take if it is to characterize successfully the linguistic knowledge which distinguishes speaker from non-speaker. Montague apparently agrees here that his theories do not suffice for this purpose (although they may succeed in characterizing part of our semantic competence, for example our competencies to determine logical consequence, ambiguity, compositionality, and so on). To do a complete job, we need to go on to define the unrelativized sense of a sentence and the unrelativized truth conditions (see also Partee, 1975). In the above passage, Montague says that, if we adopt his approach, a complete job would involve singling out one interpretation to determine the meanings of the various sentences of the language, and in addition, one world in this interpretation to determine the truth conditions for these sentences. If an interpretation can be singled out and along with it the actual world, then presumably someone who understands PTQ can use it to interpret sentences from the fragment of English PTQ addresses itself to. But this is because he brings to bear his knowledge of what the actual interpretation of his language is and of which world is the actual one. This is knowledge which PTQ does not state.

In our discussion of what we must add to an MTS theory (for instance (1)) for L in order to derive from it an absolute truth theory (for instance (2)) for L, we argued that the smallest addition would include the base clauses of an absolute truth theory for L ('N' denotes N, 'P' is true of Ps, etc.), and also a statement that truth in some interpretation A (for L) at some world p in A is absolute truth. Put somewhat differently, a semantic theory for L must state that A and *p* are the actual interpretation and world respectively. Montague, however, seems to be recommending that we pass over the articulation of the base clauses—which would essentially involve constructing an absolute truth theory for L—and instead single out the actual world and interpretation directly. Richmond Thomason, a proponent of Montague style semantics, seems to take a similar line. He agrees that Montague's theory is abstract in the sense that

it not only allows a multiplicity of interpretation assignments, but a multiplicity of interpretation structures. That is, interpretations can differ in the materials that are used to construct the space of possible denotations as well as in the particular semantic values they attach to basic expressions. (Thomason in Thomason, 1974, p. 50).

But Thomason thinks that, 'in itself, this is not damaging; one might conclude that it is merely an *empirical matter* to construct an appropriate set of entities and possible worlds for one of Montague's fragments of English' (ibid.).

Both Thomason and Montague are over-zealous about what we can reasonably expect to accomplish by appeal to MTS. In conclusion, I will argue that the route Montague and Thomason apparently opt for will not work.

First, it is not clear how to go about specifying the actual interpretation. Returning to (E) and (I): singling out the actual interpretation would involve for instance determining what function 'Barbara' denotes. But how would we specify this function? Is it the function which, given a possible world *w* as argument, has a value that 'Barbara' denotes in world *w*? The trouble with this suggestion is that, unless some further rule is laid down to deal with the notion expressed by the phrase 'what "Barbara" denotes in world *w*', we have not successfully eliminated reference to language, and therefore we will not be able to derive theorems needed to do the intellectual work we are interested in. Perhaps we can specify the actual interpretation by saying that in it "Barbara" denotes the function which, given a possible world *w* as an argument, has as value the thing which is Barbara in world *w*. What in the world does this mean? Do we really need to understand it to understand the sentence 'Barbara sekoilee?'

On the other hand, Montague and Thomason seem to be saying that, if we want to move from relativized truth conditions to absolute truth conditions, we need to single out the actual world among all possible worlds. This certainly is no easy task either. How much about a world do we need to know before we can distinguish it from all other worlds? Presumably a lot. There presumably is a class of worlds in which the number of trees in Canada is even and one in which the number is odd. So far are we from being able to single out the actual world from all others that we do not even know which class it falls in. But do we need to distinguish the actual world from all others in order to understand our language? From the point of view of PTQ and MTS theories in general, what we are seeing is that, in order to understand a language, one must have enough knowledge to single out the actual world. And this—need it be said?—is more than any speaker knows. Indeed, Montague's semantics seems, for contingent sentences, to collapse the distinction between understanding a sentence and knowing whether it is true.

Acknowledgments

The idea for this chapter derives from some comments Donald Davidson has made on model-theoretic semantics in several of his papers. I would like to thank him and John Wallace. I would also like to thank Bill Lycan, Paul Yu, and John Biro for comments on earlier drafts of this chapter.

References

Bach, Emmon (1968) Nouns and Noun Phrases. In E. Bach and R. T. Harms (eds), *Universals in Linguistic Theory*, New York: Holt Rinehart and Winston, pp. 91–124.

Cresswell, M. J. (1973) *Logics and Languages*. London: Methuen.

Cresswell, M. J. (1978) Semantic Competence. In M. Guenthner Ruetter and F. Guenthner (eds), *Meaning and Translation*, Duckworth, London, pp. 9–27.

Davidson, D. (1967) Truth and Meaning. *Synthese* 17: 304–323, reprinted in Davison (2001).

Davidson, D. (1973) In Defense of Convention T. In H. LeBlanc (ed.), *Truth Syntax and Modality*. Amsterdam: North-Holland Publishing Co., pp. 76–86, reprinted in Davison (2001).

Davidson, D. (2001). *Inquiries into Truth and Interpretation*, 2nd edn, Oxford: Oxford University Press.

Dummett, M. (1975) What Is a Theory of Meaning (pt. I)? In S. Guttenplan (ed.), *Mind and Language*, Oxford: Clarendon Press, pp. 97–38.

Field, H. (1973) Tarski's Theory of Truth. *Journal of Philosophy* 69: 347–375.

Fodor, J. A. (1975) *Language of Thought*. New York: Thomas Y. Crowell Company.

Fodor, J. D. (1977) *Semantics: Theories of Meaning in Generative Grammar*. Cambridge, MA: Harvard University Press.

Harman, G. (1974) Meaning and Semantics. In M. Munitz and P. Unger (eds), *Semantics and Philosophy*, New York: New York University Press, pp. 1–16.

Hintikka, J. (1969) Semantics for Propositional Attitudes. In J. Davis, D. Hockney, W. Wilson, (eds), *Philosophical Logic*, Dordrecht: D. Reidel, pp. 21–45.

Hintikka, J. and L. Carlson (1979) Conditionals, Generic Quantifiers and Other Applications of Subgames. In Avishai Margalit (ed.), *Meaning and Use*, Dordrecht: D. Reidel, pp. 57–92.

Jackendoff, R. (1972) *Semantic Interpretation in Generative Grammar*. Cambridge, MA: MIT Press.

Katz, J. (1966) *The Philosophy of Language*. New York: Harper.

Katz, J. (1972) *Semantic Theory*. New York: Harper.

Katz, J. (1975) Logic and Language: An Examination of Recent Criticisms of Intensionalism. In Gunderson, K. (ed.), *Language, Mind and Knowledge*, Minneapolis: University of Minnesota Press, pp. 36–130.

Katz, J. (1977) The Advantage of Semantic Theory over Predicate Calculus in the Representation of Logical Form in Natural Language. *Monist* 60: 380–405.

Katz, J., and J. A. Fodor (1963) The Structure of a Semantic Theory. *Language* 39: 170–210.

Katz, J. and R. Nagel (1974) Meaning Postulates and Semantic Theory. *Foundations of Language* 11: 311–340.

Katz, J. and P. Postal (1964) *An Integrated Theory of Linguistic Description*. Cambridge, MA: MIT Press.

Kripke, S. (1963a) Semantical Analysis of Modal Logic. I. Normal Propositional Calculi. *Zeitschrift für mathematische Logik* 9, 67–96.

Kripke, S. (1963b) Semantical Considerations on Modal Logic. *Acta Philosophica Fennica* 16, 83–94.

Lakoff, G. (1972) Linguistics and Natural Logic. In G. Harman and D. Davidson (eds), *Semantics for Natural Languages*, Dordrecht: D. Reidel, pp. 545–665.

Lepore, E. (1982) In Defense of Davidson. *Linguistics and Philosophy* 5: 277–294.

Lepore, E., and Loewer, B. (1981) Translational Semantics. *Synthese* 48, 121–133. [Chapter 1 in this volume.]

Lewis, D. (1973) *Counterfactuals*. Cambridge, MA: Harvard University Press.

Lewis, D. (1974) 'General semantics.' *Synthese* 27: 18–67.

Lyons, J. (1968) *Introduction to Theoretical Linguistics*. Cambridge, MA: Harvard University Press.

McCawley, J. (1968) The Role of Semantics in Grammar. In E. Bach and R. T. Harms (eds), *Universals in Linguistic Theory*, New York: Holt Rinehart and Winston, pp. 124–169.

Montague, R. (1974a) English as a Formal Language. In Thomason, pp. 188–222.

Montague, R. (1974b) The Proper Treatment of Quantification in Ordinary English. In Thomason, pp. 247–271.

Montague, R. (1974c) Universal Grammar. In Thomason, pp. 222–247.

Partee, B. (1975) Montague Grammar and Transformational Grammar. *Linguistic Inquiry* 6: 203–300.

Partee, B. (1977) Possible World Semantics and Linguistic Theory. *Monist* 60: 303–326.

Partee, B. (1980) Montague Grammar, Mental Representation, and Reality. In S. Ohman and S. Kanger, eds, *Philosophy and Grammar*, D. Reidel, Dordrecht (1980) 59–78. Reprinted in P. French et al., eds, *Contemporary Perspectives in the Philosophy of Language*, University of Minnesota Press, Minneapolis (1979).

Potts, T. (1975) Model Theory and Linguistics. In E. Keenan (ed.), *Formal Semantics for Natural Language*, Cambridge: Cambridge University Press, pp. 241–250.

Saarinen, E. (1977) Game-Theoretical Semantics. *Monist* 60: 406–418.

Stalnaker, R. (1972) Pragmatics. In In G. Harman and D. Davidson (eds), *Semantics for Natural Languages*, Dordrecht: D. Reidel, pp. 380–397.

Tarski, A. (1965) The Concept of Truth in Formalized Languages. In A. Tarski, *Logic, Semantics, and Metamathematics*, Oxford: Clarendon Press, pp. 152–278.

Thomason, R. (ed.) (1974) *Formal Philosophy: Selected Papers of Richard Montague*. New Haven, CT: Yale University Press.

van Fraassen, B. C. (1969) Presuppositions, Supervaluations and Free Logic. In K. Lambert (ed.), *The Logical Way of Doing Things*, New Haven, CT: Yale University Press, pp. 67–91.

Vermazen, B. (1967) Review of J. Katz, J., *The Philosophy of Language* (New York: Harper, 1966 and J. Katz, and P. Postal, *An Integrated Theory of Linguistic Description* (Cambridge, MA: MIT Press, 1964). *Synthese* 17: 350–365.

4

The Role of 'Conceptual Role Semantics'

Barry Loewer

In his 1982 paper, Gilbert Harman defends conceptual role semantics (henceforth, CRS), a theory of meaning he has been elaborating for the past decade (Harman, 1972, 1975a, 1975b). CRS is especially interesting for the way it combines issues that are central to both the philosophy of language and the philosophy of mind. According to Harman, it is founded on two claims:

[I] The meanings of linguistic expressions are determined by the contents of the concepts and thoughts they can be used to express.

[II] The contents of concepts and thoughts are determined by their 'functional role' in a person's psychology.

A corollary of [I] and [II] is that the use of symbols in calculation and thought is more basic than the use of symbols in communication. It seems to me that a good way to appreciate the role of CRS is to compare it with truth-conditional semantics (henceforth, TCS). Harman seems to consider the two approaches to be to some extent in competition with each other and has argued that TCS can make, at most, a subsidiary contribution to the theory of meaning. In my comments I will argue that the two approaches are best seen as complementary. Although CRS is a significant contribution, TCS has a central role to play in accounts of language—both language used for communication and language used for calculation and thought. In the first section, I will sketch an account of understanding language used for communication which is based on TCS. In the next section, I consider an argument of Harman's which, he thinks, demonstrates the impotence of TCS. In the third section I consider some contributions of CRS, and in the final section I argue that TCS is central to an account of language used in thought.

Truth-conditional semantics and communication

Concerning the view of Davidson (1967, 1974), Lewis (1974), and others that an account of the truth conditions of the sentences of a language is central to a theory of meaning for that language, Harman remarks:

[T]his seems wrong. Of course if you know the meaning in your language of the sentence S, and you know what the word 'true' means, then you also know something of the form 'S is true iff' [...] But this is a trivial point about the meaning of 'true' not a deep point about meaning. (1980, p. 247)

Harman goes on to grant that a theory of truth for a language may shed light on meaning, but only by specifying implications among sentences. His view is that the functional roles of logical constants 'and', 'or', 'all', and so on are mostly characterized by specifying their roles in inference. So a theory of truth can at best specify the meanings of logical constants, and does so by characterizing their conceptual roles.

I think that Harman's dismissal of truth conditional semantics is mistaken. I first want to consider an argument which shows that TCS has an important role to play in an account of communication and language understanding.

One of the goals of a theory of meaning is to characterize linguistic competence. This problem has been construed by Dummett (1975) and by Davidson (1967) as the problem of characterizing what someone must know to understand a language. Davidson's answer is, of course, that understanding a language consists, at least in part, in knowing the truth conditions of the sentences of the language. There is a simple argument, suggested by Davidson but never explicitly formulated by him, which seems to show that his answer is correct. (This argument is spelled out in more detail in Lepore and Loewer, 1981.) Consider the following communication episode. Arabella, Barbarella, and Esa are in a room, and Arabella is looking out the window. Arabella and Barbarella understand German but Esa does not. Arabella turns from the window to Barbarella and Esa and utters the words 'Es schneit'. On the basis of this utterance, Barbarella comes to believe that it's snowing (and also that Arabella believes that it's snowing), while Esa comes to believe only that Arabella said something which is probably true. We can focus on the question of what knowledge comprises Barbarella's understanding 'Es schneit' by asking what Esa would need to know to come to the same beliefs as Barbarella. The obvious candidate for such knowledge is the knowledge that 'Es schneit' is true iff it's snowing.[1] A reconstruction of the reasoning which justifies Barbarella's acquisition of the belief that it's snowing looks like this:

(1) Arabella utters the words 'Es schneit'
(2) Since 'Es schneit' is an indicative sentence and since Arabella is generally reliable, her utterance of 'Es schneit' is true,
(3) 'Es schneit' is true iff it's snowing therefore,
(4) It's snowing

---

[1] Strictly speaking, the required knowledge is that 'Es schneit' is true when uttered at time t in location w iff it's snowing at t in the vicinity of w.

Both Esa and Barbarella can come to believe that 'Es schneit' is true by knowing a bit of German grammar (enough to recognize indicative sentences) and by knowing that Arabella is reliable. But only Barbarella is in a position to go on to conclude that it's snowing, since only she understands German. And, if my argument is correct, that understanding must consist, in part, in knowing the truth conditions of the German sentence.

An objection to my argument is that understanding a language does not involve propositional knowledge, 'knowledge that'. Some, including Harman (1975b), have argued that understanding is a know-how which cannot be explicated in terms of propositional knowledge. I do not want to take up this thorny issue here, except to make one observation. Even if in our ordinary understanding of a language used for communication we do not employ propositional knowledge, I think it is clear from my example that anyone who understands such a language does know the truth conditions of its sentences.

Since there are infinitely many sentences of a language, a theory is needed to specify the knowledge in which understanding consists. As Davidson has been urging for over a decade, the kind of theory required is a Tarski-type theory of truth, since that would assign truth conditions to each indicative sentence of a language. A theory of truth will, as Harman notes, also spell out implications among sentences, but, more importantly, it will provide the core of an account of the understanding of language used in communication.

Harman's objections

Harman is unimpressed by truth conditional semantics. I think half his reasons are based on a confusion and half based on an argument which he thinks shows that TCS merely postpones the problem of providing semantics. The argument also shows why he thinks language used in thought is more basic than language used in communication. First, the confusion. Harman says that ' "Snow is white" is true iff snow is white' expresses a trivial point about the meaning of 'true', not a deep point about the meaning of 'Snow is white'. I suspect that the reason why he thinks this is that anyone who is familiar with quotations and with the disquotational effect of 'is true' can recognize that ' "Snow is white" is true iff snow is white' expresses a truth, even if they do not have the slightest idea what 'Snow is white' means. This is correct but irrelevant. A theory of truth employed as an account of understanding does not attribute the knowledge that the sentence ' "Snow is white" is true iff snow is white' is true to a speaker, but the knowledge that 'Snow is white' is true iff snow is white. There is a world of differ ence between the two knowledge attributions. The second, but not the first, justifies the inference from a belief that the utterance 'Snow is white' is true to a belief that snow is white. It is very easy to make the mistake of thinking that sentences like ' "Snow is white" is true iff snow is white' are trivial when the sentence embedded in it (in quotation marks) is in a language you understand. But the reason is that you exploit your understanding of the language to recognize their truth. The truth conditional semanticist is in fact claiming that your understanding 'Snow is white' partially consists in your knowing its truth conditions.

The argument that underlies Harman's dismissal of truth-conditional semantics is formulated in another paper:

Davidson would [presumably] say that the speaker understands [the sentence 'Snow is white'] by virtue of the fact that he knows it is true if and only if snow is white. The difficulty [...] is that [for the speaker to know this he] needs some way to represent to himself snow's being white. If the relevant speaker uses the words 'Snow is white' to represent in the relevant way that snow is white [...] Davidson's (theory) would be circular. And, if speakers have available a form of Mentalese in which they can represent that snow is white, so the (theory avoids) circularity, there is still the problem of meaning for Mentalese. (Harman, 1975a, p. 286)

Harman's argument can be reconstructed as follows:

(1) Suppose that to understand 'Snow is white' is to know that 'Snow is white' is true iff snow is white.
(2) But, to know that S, one must have a way of representing S to oneself.
(3) That is, to know that S, there must be in one's language of thought a token which means that S.
(4) If the language of thought is English, then Davidson's theory is circular.
(5) If the language of thought is Mentalese, then Davidson's theory is incomplete, since it fails to specify semantics for Mentalese.

Harman seems to take this argument as establishing his view that TCS makes no contribution (other than helping to characterize the functional role of certain words) to the theory of meaning. But so strong a conclusion is not supported by the argument. I have previously argued that understanding a language used for communication involves knowing the truth conditions of its sentences, and nothing in Harman's argument undermines this claim. Neither does Harman's argument show that sentences in a language of thought, if there is such a language, do not have truth conditions, or that characterizing their truth conditions is not an important part of characterizing their meanings. What his argument does show is that, if we grant the truth of (3), then understanding a language of thought cannot be explained as *knowing* the truth conditions of its sentences. To attempt to do so would involve one in a regress, since we would then have to explain knowledge of the truth conditions of a language of thought by postulating another language of thought, understood by the knower. This is an important point. It does show that, if (3) is correct, then TCS is not the whole story concerning meaning. (Similar points are made by Field, 1978.)

Premise (3) of the argument is crucial. Sometimes the view that to believe that *p* is to be related to a token of some language of thought (English or Mentalese) which means that *p* seems to be taken to be a conceptual truth. But this is a mistake. It is an empirical hypothesis—perhaps an especially promising one, but still an hypothesis, which may ultimately prove to be false. If it is rejected, then Harman's argument contra Davidson collapses. TCS for a language used in communication may be all we ever get or need by way of a theory of meaning. Since the language-of-thought hypothesis is empirical, the exact nature of the language (or languages) of thought remains to be specified by the development of cognitive psychology. This poses two possible difficulties for Harman's view, embodied in his principles [I] and [II]. First, it may turn out that the language of thought postulated by cognitive psychology does not support [I], that the meanings of linguistic expression are determined by the contents of the thoughts which they are used to express. It could turn out that sentences in Mentalese underdetermine the meanings of public language. I will discuss this possibility in the final section. Second, it may turn out that, in characterizing the language of thought, cognitive psychology will employ, as primitive, a notion of linguistic meaning. I am not sure how this might happen, but Davidson (1975) has given some arguments which seem to point in this direction. He has argued there that 'a creature must be a member of a speech community if it is to have the concept of a belief' and that '[s]omeone cannot have a belief unless he understands the possibility of being mistaken, and this requires grasping the contrast between truth and error— true belief and false belief'. (p. 170). If Davidson is correct, then an account of belief and, if (3) is correct, of the language of thought will presuppose concepts of linguistic meaning. This is not the place to try to provide an assessment of Davidson's argument. I mention it only to make the point that Harman's views that language used in thought is more basic than language used in communication and that truth-conditional semantics must be based on non-truth-conditional semantics for the language of thought are not inevitable.

What is conceptual role?

Harman conceives of CRS as providing an answer to the question: what makes something a concept with the content C? For example, he writes: 'What makes something the concept red is in part the way the concept is involved in the perception of red objects in the external world' (1980, p. 247). CRS's answer to this question is that an expression of a particular person P's Mentalese is a concept with content C in virtue of playing a certain role in P's psychology. There are two questions I want to consider about this view:

(1) What is it to specify the conceptual role of an expression of P's Mentalese?
(2) How are the conceptual roles of expressions of Mentalese related to reference and truth conditions?

CRS can be developed either atomistically or holistically. Procedural semantics, fashionable within much work in the Artificial Intelligence community, is an example of the former approach (see Suppes, 1980). According to procedural semantics there is a stock of primitive expressions and various devices for constructing complex expressions. Associated with the former are procedures for determining whether or not an individual (presented in some canonical way) falls under the extension of the expression. Associated with the latter are algorithms for constructing complex procedures. Fodor (1981b)has pointed out affinities between procedural semantics and 'old-fashioned verificationism', and argued that the well-known ills of the latter are inherited by the former. This does not seem to affect Harman's account, since it is clear that he favours the holistic approach. However, he says very little about how the details of CRS are to be developed. The only detailed model of CRS which has been suggested is Hartry Field's probabilistic semantics (see Field, 1977). But Harman rejects Field's account because it is unrealistic 'since keeping track of probabilities involves memory and calculating capacities which are exponentially exploding functions of the number of logically unrelated propositions involved' (1981, p. 247).

Perhaps Harman does not spell out the details of CRS because he correctly believes that CRS is an empirical hypothesis whose exact form awaits developments in cognitive psychology. However, I think he does suppose that CRS will have the following general form. CRS will be a component of a functionalist theory of mind. A functionalist psychological theory contains a specification of possible inputs (experience), a specification of possible outputs (behavior), a specification of mental states, and various lawful statements (perhaps probabilistic laws) linking them. According to the language-of-thought hypothesis, a specification of mental states, or a subset of the states, requires supposing that they have a syntax and semantics. Now the central claim of CRS is that the content of an expression of the language of thought has been specified when one specifies all of this theory. Harman's version is holistic since specifying the content of an expression involves spelling out its connections and non-connections with all other expressions as well as with inputs and outputs.

Despite the sketchiness of my (and Harman's) account of CRS, we can see that it does fill certain lacunae left by TCS. We saw in the previous section that, assuming the language-of-thought account of belief, Harman's argument shows that TCS cannot explain what it is to understand Mentalese. The question of how one understands the language one thinks in does seem to be a peculiar one. Fodor has remarked that it doesn't even make sense. CRS clarifies the situation. It is plausible that understanding a certain concept involves being able to use that concept appropriately. For example, to understand the concept of 'red' is, in part, to be able to discriminate red things. According to CRS, an expression in P's Mentalese has the content of the concept 'red' just in case this concept plays the appropriate role in P's psychology, including his discriminating red things. It follows that, if some expression of P's Mentalese is the concept 'red', then P automatically understands it. The answer may appear to be a bit trivial—P understands the expression of his Mentalese since, if he didn't, it wouldn't be his Mentalese—but it is the correct answer. If there are any doubts, compare the question we have been considering with: 'in virtue of what does a computer 'understand' the language it computes in?' Of course the understanding involved in understanding Mentalese is different from the understanding one has of a public language. I argued that understanding the latter involves knowing truth conditions. Not only would knowledge of truth conditions contribute nothing to explaining how we understand Mentalese but, it is clear, we do not know the truth conditions of Mentalese sentences. (Or, for that matter, even the syntax of Mentalese.) If P were to encounter a sentence of Mentalese written on the wall (in contrast to its being in just the right place in his brain), P wouldn't have the vaguest idea as to what it means, because P does not know its truth conditions.

There are certain sentences which seem to create difficulties for the truth-conditional account of understanding when it is coupled with the language-of-thought account of belief. Suppose that 'Hesperus' and 'Phosphorus' were directly referential—that is, that the semantic value of these two names were simply the planet Venus. Then the sentences 'Hesperus is too hot to sustain life' and 'Phosphorus is too hot to sustain life' would have identical truth conditions. If there is a language of thought, then the truth conditions of these sentences are represented by some sentence S of Mentalese. It would seem to follow that anyone who understands both sentences and believes that 'Hesperus is too hot to sustain life' is true will also believe, if she follows the consequences of her beliefs, that 'Phosphorus is too hot to sustain life' is true. She should be as willing to assert one as the other. But this is certainly wrong. The obvious way out is to suppose that the two sentences express beliefs which have different representations in Mentalese. But then there must be more to the semantics of Mentalese sentences than their truth conditions. According to CRS, the two representations may differ in conceptual role, even if they have the same truth conditions. It might be part of the conceptual role of the thought that Hesperus is too hot to sustain life to tend to produce, under certain conditions, utterances of 'Hesperus is too hot to sustain life', but not utterances of 'Phosphorus is too hot to sustain life'. If this is correct, then, even if one holds that truth-conditional semantics provides the core of an account of language understanding, it will need to be supplemented with an account of conceptual roles.

The indispensability of truth conditions

In this section I will discuss in a bit more detail the relations between CRS and TCS. Harman's view expressed in [I] and [II] is that the truth conditions of an indicative sentence are determined by its conceptual role. As I have already mentioned, he also considers sentences which express truth conditions to be trivial and does not see a theory which specifies truth conditions as making much of a contribution to the theory of meaning. The issues involved in the relations between CRS and TCS come to the fore when one considers Putnam's (1978) twin-earth stories.

Putnam asks us to imagine a twin Earth which is just like the Earth, except that what they call 'water' there is a liquid whose molecular structure is XYZ instead of H_2O. According to Putnam, the reference of 'water' in English is H_2O, while the reference of 'water' in Twin-English is XYZ. He uses this example to argue that meaning cannot be both what is in the head and what determines reference. Presumably the representation of H_2O in an earthling's head and the representation of XYZ in his Twin Earth doppleganger's head are of the same type, since the two are, physically, type-identical. Since the meanings of these representations differ, they do not determine reference. If meaning does determine reference, it is not in the head.

One way of reacting to Putnam's example is to deny that their 'water' and ours differ in reference. But Harman accepts this intuition of Putnam's. The example seems to create difficulties for CRS because, on natural understandings of what is to count as a conceptual role, it would seem that the conceptual roles of the earthling's concept of 'water' and his doppleganger's concept of 'water' are the same but their contents are different, thus violating claim [II].

So far as I can see there are two ways out of the Twin-Earth problem for an advocate of CRS who accepts Putnam's intuitions about reference. The first way is to include, in the psychological theory that characterizes an organism's Mentalese, microphysical descriptions of possible inputs. Then the content of a concept would be specified by specifying the theory and the inputs the organism actually experienced. The content of the concepts of 'water' on Earth and Twin Earth would be different, since the respective concepts developed in the two organisms in response to different kinds of inputs: H_2O inputs on Earth, XYZ inputs on Twin-Earth.

This response to the problem does not seem very attractive from the perspective of CRS. As I previously urged, the language-of-thought hypothesis is an empirical psychological hypothesis. If Fodor is correct in his claim that psychological theory will be 'methodologically solipsistic' (1981a), then the theory developed by psychologists will not characterize inputs in terms of their microphysical structures. Methodological solipsism may turn out to be mistaken, but it is difficult to see its rejection by CRS in response to the Putnam problem as anything other than an ad hoc manoeuver to try to save CRS.

In any case, Harman's response (1981, p. 248) is a different one. He revises his position by replacing [II] by the principle that the content of a concept is determined by its functional role *relative to some normal context*. Even though the conceptual roles of the concept of 'water' are the same on Twin Earth and Earth, their contents differ, since contents are evaluated relative to normal context, and the normal contexts on earth and on twin-earth differ. This revision is an appropriate one. In any case, context needs to be taken into account to handle thoughts that we express using indexicals. However, this revision may be more far-reaching than Harman realizes.

Among the features of context which may need to be specified in order to fix the content of a thought, there are, at least, the thinker, the time of the thought, the references of demonstrative elements in the thought, and, if Putnam is correct, certain microphysical facts about the thinker's 'normal' environment. But it might turn out that context has an even heavier burden to bear. It will depend on how the psychological theory which characterizes the language of thought specifies inputs and outputs. If inputs are specified as, say, patterns of neural activity in the optic nerve, then it might very well be that, in order to fix the content of a sentence of Mentalese, we will need a great deal of information about the physical objects which caused particular patterns of neural activity in the optic nerve.

By revising [II] to accommodate Putnam's example, Harman is recognizing that Mentalese sentences have truth conditions. Given a complete psychological theory for P, the conceptual roles of P's Mentalese sentences are characterized. Conceptual role plus context yields truth conditions. But, from the perspective of CRS, it is difficult to discern why we should be interested in truth conditions at all. Harman seems to think of truth conditions as epiphenomenal, the by-product of possessing the concept of truth and quotation, but not as themselves making any contribution to semantics. But this distorts the relation between conceptual role and truth conditions. As long as we confine our interests to explaining the behaviour of an organism for which we possess a psychological theory of the sort Harman hopes for, we will not need to employ the truth conditions of its Mentalese sentences, or even the fact that they have truth conditions. But this does not show that truth conditions have no semantic function. I will give two reasons for believing that they are essential.

First, there are certain relations between a believing organism and its environment which involve truth conditions of the organism's beliefs. For example, if there is a fox in the vicinity, then a rabbit in good working order is likely to come to believe that there is a fox in the vicinity. It is not unusual to explain why it is that someone believes that q by pointing out that q is the case and that he is appropriately situated with respect to the situation q. If S is the sentence of P's Mentalese that he is related to when he believes that q, then the truth conditions of S—namely, S is true iff q—can play a part in an explanation of P's acquisition of the belief that q. By neglecting truth conditions, CRS misses an important feature of belief states: that they are information-bearing states.[2] Under suitable conditions, P's believing that q carries the information that q. This feature of belief is intimately related to the account of communication and understanding sketched in the first section. Arabella's assertions about the weather are reliable because she asserts what she believes and her beliefs carry information about the weather. Truth conditions enter the picture because her belief that it's snowing carries the information that it's snowing in virtue of its having the truth conditions that it's snowing. An account of why it is that Barbarella believes that it's snowing involves

---

[2] Dretske, 1980 develops an information-based account of belief. I criticize his account in Loewer 1987.

not only the truth conditions of Arabella's utterance 'Es schneit', but also the truth conditions of her belief state.

Suppose that we have a psychological theory of the rabbit mentioned a paragraph back, i.e. the sort of theory that Harman envisions as characterizing the rabbit's mental states. Assuming that the rabbit has beliefs and that the language-of-thought hypothesis holds for them, the theory will provide CRS for Rabbitese. We can use this theory to explain why the rabbit bears the belief relation to certain sentences of Rabbitese, for example the sentence we translate by 'There is a fox in the vicinity'. The explanation would cite the rabbit's mental states at some time previous to its acquiring the belief and certain inputs to its sensory system. Given this information, the psychological theory will imply, or perhaps only make probable, that the rabbit acquires the belief that there is a fox in the vicinity. How does this explanation compare with the one that employs the truth conditions of the belief that there is a fox in the vicinity and cites the fact that there is a fox in the vicinity? On the positive side, it is much more specific and informative. The truth-conditional explanation is, at best, a rough one, since rabbits don't invariably believe that foxes are present when they are. On the negative side, it is a much more complicated explanation. For most purposes the gain in precision probably will not offset the loss due to complexity. In any case the two explanations are not in competition. They are related to each other in something like the way in which a higher-level functional psychological explanation is related to a lower-level neurophysiological explanation. In our example, the lower-level psychological account helps to explain how the higher-level account that appeals to truth conditions works. But even if we had a complete psychological theory of rabbits, it would not render the higher-level explanation useless or redundant. We would still need it to understand why the rabbit believes what it does.

My second reason for saying that truth-conditional semantics is central to a semantic account of Mentalese is a development of the first reason. It is no accident that generalizations connecting situations in the world with an organism's beliefs via those beliefs' truth conditions hold. Rabbits and other believers are constructed so that, when they are in their proper environments, they acquire mostly true beliefs. Daniel Dennett, in a number of papers,[3] has distiguished between being a believer and believing truths, and forcefully argued that 'true believers mainly believe truths'. If this is correct, then the fact that beliefs, or sentences of Mentalese, have truth conditions is central to our notion of belief.

I am not sure that Harman would disagree very much with this. He never denies that Mentalese sentences have truth conditions. But in the course of his advocacy of CRS he does claim that the possession of truth conditions is a rather trivial matter. In my comments I have endeavoured to show that, on the contrary, truth conditions are at the very heart of semantical theory.

---

[3] See Dennett, 1981, and 1987, chapter 2.

Acknowledgments

I would like to thank Robert Laddaga and Ernie Lepore for helpful discussions on the issues surrounding conceptual role semantics.

References

Davidson, D. (1967) Truth and Meaning. *Synthese* 17 (1967): 304–323; reprinted in Davidson (2001).

Davidson, D. (1974) Belief and the Basis of Meaning. *Synthese* 27: 309–323; reprinted in Davison (2001).

Davidson, D. (1975) Thought and Talk. In S. Guttenplan (ed.), *Mind and Language*, Oxford: Oxford University Press, pp 723; reprinted in Davidson (2001).

Davidson, D. (2001) *Inquiries into Truth and Interpretation*, 2nd edn, Oxford: Oxford University Press.

Dennett, D. (1981) Three Kinds of Intentional Psychology. In R. A. Healey (ed.), *Reduction, Time and Reality*, Cambridge, UK: Cambridge University Press, pp. 37–61.

Dennett, D. (1987) *The Intentional Stance*, Cambridge, MA: The MIT Press.

Dretske, F. (1980) *Knowledge and the Flow of Information*. Cambridge, MA: MIT Press.

Dummett, M. (1975) What is a Theory of Meaning, Part One. In S. Guttenplan (ed.), *Mind and Language*, Oxford: Oxford University Press, pp. 97–138.

Field, H. (1977) Logic, Meaning, and Conceptual Role. *Journal of Philosophy* 74: 379–408.

Field, H. (1978) Mental Representations. *Erkenntnis* 13, pp. 9–61.

Fodor, J. (1981a) Methodological Solipsism. In J. Fodor, *Representations*. Cambridge, MA: MIT Press, pp. 225–257.

Fodor, J. (1981b) Tom Swift and his Procedural Grandmother. In J. Fodor, *Representations*, Cambridge, MA: MIT Press, pp. 204–225.

Harman, G. (1974) Meaning and Semantics. In M. Munitz and P. Unger (eds), *Semantics and Philosophy*, New York: New York University Press, pp. 1–16.

Harman, G. (1975a) Language, Thought, and Communication. In K. Gunderson (ed.), *Language, Mind, and Knowledge*, Minneapolis: University of Minnesota Press, pp. 270–298.

Harman, G. (1975b) *Thought*. Princeton, NJ: Princeton University Press.

Harman, G. (1982) Conceptual Role Semantics. *Notre Dame Journal of Formal Logic* 23: 242–257.

Lepore, E. and B. Loewer (1981) Translational Semantics. *Synthese* 48: 121–133. [Chapter 1 this volume].

Lewis, D. (1974) General semantics. *Synthese* 27: 18–67.

Loewer, B. (1987) From Information to Intentionality. *Synthese* 70: 287–317.

Putnam, H. (1975) The Meaning of 'Meaning.' In K. Gunderson (ed.), *Minnesota Studies in the Philosophy of Science*, Vol. 7, Minneapolis: University of Minnesota Press, pp. 131–193.

Suppes, P. (1980) Procedural Semantics. In R. Haller and W. Grassl (eds), *Language, Logic, and Philosophy*, Proceedings of the 4th International Wittgenstein Symposium, Kirchberg am Wechsel, Austria, Vienna: Hölder-Pichler-Tempsky, pp. 27–35.

5

Dual-Aspect Semantics

Ernest Lepore and Barry Loewer

Frege's notion of sense plays (at least) two roles in his theory of meaning. One role concerns the relations between language and reality: an expression's sense determines its reference. The other role relates a language to the mind of someone who understands it: to understand an expression is to grasp its sense. The dual role of sense is seen clearly in Frege's account of the semantics of identity statements. 'The morning star = the evening star' is true, since the sense of 'the morning star' and the sense of 'the evening star' determine the same reference. The sentence is cognitively significant since it is possible for someone to know the senses expressed by the expressions 'the morning star' and 'the evening star', yet not to know that they determine the same reference.

During the last fifteen years or so there has been a sustained attack on the Fregean conception of sense. An examination of proper names, indexicals, and natural kind terms has led many philosophers of language and mind to conclude that no single notion of sense can play both roles. Hilary Putnam puts the point by saying that 'no theory can make it the case that "meanings" are in the head and simultaneously make it the case that "meanings" determine external world reference' (Putnam, MH, p. 12). Of course, for Frege, meanings—that is, senses—are in the head (in that they are grasped) and determine reference.

A number of philosophers have responded to these arguments by constructing two-tiered or dual aspect theories of meaning. We will call them DATs (see Block, 1986; Field, 1977; Fodor, 1980; Harman, 1973, 1974, 1982; Loar, 1981; Lycan, 1981, 1982, 1984; McGinn, 1982). According to these accounts, a theory of meaning for a language L consists of two distinct components. One component is intended to provide an account of the relations between language and the world: truth, reference, satisfaction, and so on. The other is supposed to provide an account of understanding and cognitive significance. In this chapter we will examine a particular proposal concerning the appropriate form of a DAT according to which the two components are:

(i) a theory of reference and truth for L;
(ii) a characterization of the conceptual roles of sentences and other expressions of L.

We will contrast DATs with an approach which is like Frege's in one important respect: it employs a single notion designed to serve both the purpose of the theory of reference and that of the theory of understanding. Its central tenet is that a theory of meaning for L is a certain kind of truth theory for L. Since Donald Davidson is the most prominent and subtle defender of this approach, we will call such theories of meaning 'Davidsonian truth theories' (Davidson, 1967, 1973b, 1974; Lepore, 1982b, 1983; Loewer, 1982; Lepore & Loewer, 1981, 1983). At first it may seem that this approach is contained in a DAT, since the latter has a truth theory as one of its components. But this is not so. We will show that the truth theory component of a DAT is quite different from that of a Davidsonian truth theory. We will argue that, by separating a theory of meaning into a theory of reference and a theory of conceptual role, DATs are unable to serve as theories of interpretation or as accounts of cognitive significance.

The organization of our chapter is as follows. We first examine the problems that motivate the construction of DATs. Then we discuss the form of a DAT, focusing primarily on a proposal due to Colin McGinn (1982). Next we develop a Davidsonian theory of meaning, showing how truth theories serve as theories of interpretation. As such, they provide both an account of truth and reference, on the one hand, and an account of understanding and cognitive significance, on the other. In the next section we show that DATs do not make adequate theories of interpretation. We also challenge their adequacy as theories of meaning for languages of thought. Finally, we return to the problems motivating DATs and we discuss the extent to which they can be accommodated within a Davidsonian framework.

Motivation for DAT

Hilary Putnam asks us to imagine two planets, Earth and Twin Earth, and two of their residents, say, Arabella and Twin Arabella. Twin Earth is almost a physical replica of Earth. The only difference is that, on Twin Earth, the clear liquid the Twin Earth people drink, which fills their oceans, and which they call 'water', is composed not of H_2O molecules but of XYZ molecules (Putnam, 1975).

According to Putnam, the expression 'water' on Earth refers to the stuff composed of H_2O and not to the one composed of XYZ. It is exactly the reverse for the expression 'water' on Twin Earth. This is so even if no speakers of English and Twin English know the molecular structures of water and twin water, or can distinguish between the two. Putnam argues as follows: in Frege's theory, to understand an expression, say, 'water', is to 'grasp' its sense. Exactly what it is to grasp a sense is not all that clear, but it is to be in some psychological state or other; perhaps the state of believing that 'water' expresses a certain sense. Since Arabella and Twin Arabella are physically type-identical,[1] they are in type-identical psychological states. So, if each

---

[1] Except for XYZ replacing H_2O in Twin Arabella's body. But this, of course, is irrelevant to the argument. Putnam distinguishes 'narrow psychological states' from 'broad psychological states'. Only the former supervene on brain states described in neurophysiological terms. Clearly Putnam supposes that 'grasping a sense' is a narrow psychological state.

understands her word 'water', then each grasps the same sense. But the references of their words differ. Putnam concludes that, if sense is what is grasped when understanding an expression, then sense does not determine reference. If sense is what determines reference, then sense is not what is grasped in understanding.

Putnam's initial response this argument was to distinguish two components of meaning. One he calls 'stereotype'. It is the information which linguistically competent speakers associate with an expression. The stereotype of 'water', as used both on Earth and on Twin Earth, consists in the information that water is a clear liquid, which quenches thirst, which fills the oceans, and so on. To understand 'water' is to know its grammatical role and its stereotype. This is supposed to be the 'mind' component of meaning (Putnam, 1975).

The second component of meaning is reference. On Putnam's account, the reference of a natural-kind expression like 'water' is determined by facts which are outside the minds of users of the expression. For example, 'water' refers to whatever stuff is structurally similar to this stuff (pointing at samples of water). Given that water is H_2O (and that the relevant kind of structural similarity is sameness of chemical composition), the extension of 'water' on Earth is H_2O. Analogously, the extension of 'water' on Twin Earth is XYZ. On Putnam's theory, the stereotype of an expression is the mind component of sense, its reference is the world component and, as the Twin Earth story shows, the first does not determine the second. The theory of meaning thus divides into two parts: a theory of understanding (and cognitive significance) and a theory of reference (and truth).

Putnam's argument for the bifurcation of meaning depends on accepting his view that 'water' on Earth refers to H_2O, while 'water' on Twin Earth refers to XYZ. This is a claim which can be (and has been) disputed (Zemach, 1976). But, even if Putnam is mistaken about the semantics of natural-kind terms, there are other examples that lead to dual-component views.

Imagine that Arabella and Twin Arabella, each, utter 'I am thirty years old'. Once again, they are in identical psychological states, but the references of their utterances of 'I' differ, and even the truth values of their utterances can differ. They are physical replicas, but Arabella came into existence only a few minutes ago. Arabella's and Twin Arabella's understanding of 'I' are the same, although their references and the truth values of their utterances differ. This shows that, if the sense of 'I' is what is grasped by a person who understands 'I', then that sense does not by itself determine reference. On the other hand, if sense determines reference, then Arabella and Twin Arabella do not grasp the same sense. David Kaplan, among others (Kaplan, 1989; see also Perry, 1978, 1979; White, 1982) distinguishes the character of an expression from its content in a context. The character of an expression is a function from contexts of utterances to contents; for example the character of 'I', according to Kaplan, maps a context of utterance onto the utterer. It is the character of 'I am thirty years old' that is grasped by someone who understands the sentence. The utterance's content is its truth conditions. When Arabella and Twin Arbarella, each, utter 'I am thirty years old', what is in their minds may be the same (they have the same understanding of the sentence), but their utterances have different contents and so may differ in truth value. Kaplan's account, like Putnam's, is a two-tiered theory of meaning. But it differs from Putnam's in that stereotype is unlike character: it does not determine content relative to context, at least as 'context' of utterance is normally construed (see White, 1982).

Our discussion so far seems to show that two expressions can have the same stereotype, or character (or whatever corresponds to cognitive significance), and yet possess tokens which differ with respect to reference and truth conditions. It has been argued, conversely, that sentences with the same truth conditions can differ with respect to the understanding component of meaning (Kripke, 1979). According to Kripke (1972), proper names designate rigidly. Many have been taken this to imply that the truth conditions of the sentences 'Cicero is bald' and 'Tully is bald' are identical. But what are we to make of Arabella, who understands both sentences and assents to the first but dissents from the second? If understanding a sentence is knowing its truth conditions, then it follows that Arabella is flatly contradicting herself, since she is asserting and denying statements with identical truth conditions. Furthermore, it seems to follow that she has contradictory beliefs. She believes that Cicero is bald and she believes that Cicero is not bald. But these are not ordinary contradictions, since no amount of thought on her part would enable her to recognize that she has contradictory beliefs. William Lycan (1982; see also Lycan, 1985, pp. 90–91) reacts to this problem by saying:

Nothing that [Arabella] carries in her head enables her to tell that [...] 'Cicero' and 'Tully' represent the same person. And, therefore, there is no way for her to deduce from her mental machinery anything she could recognize as a contradiction. The names 'Cicero' and 'Tully' obviously play distinct computational roles for [Arabella]. (Lycan, 1982, p. 26)

Lycan intends this as a solution to the problem as it arises for mental representations. His idea is that there are two distinct ways of semantically individuating Arabella's mental representations. One according to truth conditions, and another according to computational role. If we semantically evaluate with respect to truth conditions, Arabella believes that Cicero is bald and also believes that Cicero is not bald. But this does not impugn her rationality, and thus it would seem that truth conditions are not part of Arabella's 'mental machinery'. If we semantically evaluate with respect to computational role, the beliefs that Cicero is bald and that Tully is bald will be distinct, since the representations 'Cicero is bald' and 'Tully is bald' have different computational roles for Arabella. When her beliefs are individuated in terms of computational role, Arabella does not have contradictory beliefs.

We could extend Lycan's account (though he does not make this extension) to the semantics of natural languages, if we could find something to play the part of computational role for natural language expressions. The simplest suggestion is that the computational role of a person's sentence S at time *t* is the same as the computational role of the mental representation constituent of the belief expressed by S for P at t. 'Cicero is bald' and 'Tully is bald' have the same truth conditions in English, but they may differ in their computational roles for a particular speaker at t. It is computational role that characterizes one's understanding of an expression. We can see how a person might understand both 'Cicero is bald' and 'Tully is bald', and might assert one and deny the other, even though the expressions have the same truth conditions.

As our discussion of Lycan's proposal makes clear, DATs have been proposed for mental representations, a.k.a. languages of thought, as well as for natural languages. Jerry Fodor, the principal proponent of languages of thought, has been developing a theory of mental states and processes he calls the computational theory of mind (henceforth, CTM). According to CTM, mental states and processes are computations over representations. For example, believing that snow is white is being in a certain computational relation to a representation which means that snow is white. The system of mental representations is like a language in that representations possess both a syntax and a semantics. It is a central tenet of CTM that computations apply to representations in virtue of their syntactic features (Fodor, 1980, p. 226). While Fodor admits that it is not all that clear what count as syntactic features, he is clear that semantic properties, for example truth and reference, are not syntactic. The mind (and its components) has no way of recognizing the reference or truth conditions of the representations it operates on. Instead, it operates on syntactic features of representations which 'represent' the semantic features. The computational role of a mental representation must depend upon, and only upon, those properties of representations which do not advert to matters outside the agent's head (McGinn, 1982, p. 208).

Fodor claims that a consequence of CTM is a formality condition, which specifies that in CTM psychological states count as different states only if they differ computationally. Applied to belief, this means that beliefs can differ in content only if they contain formally distinct representations. This supervenience principle, that S and S* are distinct psychological states only if they are distinct computationally, lies at the heart of CTM. Although Fodor endorses the formality condition, he also thinks that cognitive psychology contains true generalizations connecting propositional attitudes with each other, with environmental conditions, and with behaviour. An example of the sort of generalization he has in mind is this: if someone wants to go downtown and believes that the bus provides the only way to get there, then, *cateris paribus*, he will take the bus. As Fodor emphasizes, the specification of propositional contents in these generalizations is essential to their explanatory role. It is a person's belief *that the bus provides the only way to get downtown* that explains that person's taking the bus. At first, this may seem incompatible with the claim that only formal properties of representations are relevant to the computations which produce behavior. However, there is no incompatibility as long as the contents of attitudes are specified in a way that respects the formality condition. This means that two representations can differ in content only if they differ syntactically. Fodor observes that a characterization of meaning which conforms to the formality condition is *methodologically solipsistic* (see Putnam, 1975), in that differences of meaning depend entirely upon internal mental characteristics—for example computations over representations.

We have described Fodor's views at some length because we want to show why a DAT theory seems to fit the bill as a theory of meaning for languages of thought. Fodor observes that truth-conditional semantics for a language of thought is not methodologically solipsistic. It fails to conform to the formality condition (Fodor, 1982, p. 22). Putnam's Twin Earth examples show this. Arabella and Twin Arabella are computationally identical when each one is thinking what each one would express by uttering 'water is wet'. So each one bears the same computational relation to formally identical representations. But the truth condition of the token representation in Arabella's mind is that H_2O is wet, while the truth condition of the representation in Twin Arabella's mind is that XYZ is wet. There is a difference in truth conditions, without a corresponding difference in formal properties. The characterization of contents in terms of truth conditions may seem defective from the perspective of CTM in another way as well. 'Water is wet' and 'H_2O is wet' are claimed to have the same truth conditions, but certainly there is a difference between believing that water is wet and believing that H_2O is wet. Truth conditions seems to be both too fine-grained (the Twin Earth problems) and too coarse-grained (as in Kripke's puzzle, 1979) to specify the contents of mental representations.

It should be clear why DATs have been proposed as theories of meaning for languages of thought. The truth-conditional component of a DAT characterizes the relations between representations and the world. But a second component is needed, which characterizes content in a way that conforms to the formality condition and is fine-grained enough to capture differences in belief like the one mentioned in the previous paragraph. This second component is the mind-component aspect.

The form of DAT

So far we have discussed some issues which motivate a distinction between two aspects of meaning. Some philosophers have claimed that the correct way for a theory of meaning to accommodate the two aspects is by containing two *autonomous* components: a truth-conditional component and a component accounting for the use or understanding features of meaning. Colin McGinn (1982, p. 229) explicitly advocates such a view:

For perspicuity we can separate out the two contributions by taking the meaning ascription as equivalent to a conjunction: For S to mean that *p* is for S to be true iff Q for some 'Q' having the same truth conditions as '*p*', and for S to have some cognitive role ϕ such that '*p*' also has the cognitive role ϕ.[. . .] Now to have a complete theory of meaning would be to have adequate theories corresponding to each conjunct of this schema.

McGinn is not only claiming that an adequate theory of meaning consists of two separate theories, but is also offering an analysis of 'S means that p'.

The first component, the truth theory, may seem relatively unproblematic. it is supposed to entail, for each sentence S of language L, an instance of:

(T) S is true in L iff p,

where 'S' is replaced by a structural description of a sentence of L and 'p' is replaced by a metalanguage sentence which specifies S's truth conditions. Tarski (1956) required that the sentence replacing 'p' be a *translation* of the sentence replacing 'S'. Putnam, Field, Fodor, and McGinn do not have this conception in mind. Fodor (NC, p. 9) says that 'a truth condition is an actual or possible state of affairs. If S is the truth condition of (the formula) F, then F is true iff S is actual'. According to McGinn (1982, p. 232), 'a truth theory is a specification of the *facts* stated by sentences of the object language, in the intuitive sense of that recalcitrant notion'. As McGinn says, 'fact' is a recalcitrant notion. 'State of affairs' is no clearer. However, it is clear that some advocates of DATs would count 'water is wet' (uttered by an English speaker) as stating the same fact or state of affairs as 'H_2O is wet', and 'Tully is bald' as stating the same fact or state of affairs as 'Cicero is bald'. So their characterization of an adequate truth theory is different from Tarski's, since two sentences can state the same fact (or state of affairs) without being good translations of each other. Of course, this is not a worry for advocates of DATs, since sameness of meaning requires not only sameness of truth conditions but sameness of cognitive role as well.

The second component of McGinn's DAT is a theory of cognitive role. Other authors use the phrases 'conceptual role' (which we prefer) and 'computational role' for similar, though perhaps not identical, notions. While the idea of conceptual role has been around for a while, the form of a theory of conceptual role is much less clear than the form of a truth theory. Sellars speaks of two sentences having the same conceptual role if they are related by inference, both deductive and inductive, to the same sentences in the same ways. Sellars also includes relations between sentences and perception and action, 'language entry and exit rules', in his specification of conceptual role (Sellars, 1956, 1963, 1969). Harman, thinking of the language of thought, characterizes the conceptual role of an expression by its relations to perception, to other expressions, and to behaviour (Harman, 1973, 1974, 1982). Both Sellar's and Harman's characterizations suggest that conceptual role theories for a language L will take the form of a theory of inference for L, combined with a causal theory of perceptual inputs and outputs. But neither provides detailed accounts of these theories.

McGinn relies on Hartry Field's account of conceptual role (Field, 1977). Field characterizes conceptual role in terms of a probability function defined over all the sentences of a person's language. It specifies a person's commitments concerning how she will change her degrees of belief when she acquires new information. The probability function, by specifying inductive and deductive relations, characterizes the conceptual roles of expressions. A and B are said to have the same conceptual role if and only if $P(A/C) = P(B/C)$ for all sentences C in the language. On this account, 'Tully is bald' and 'Cicero is bald' may have different conceptual roles for a person, since there may be an S for which P ('Tully is bald'/S) P ('Cicero is bald'/S). The conceptual role of a non-sentential expression is specified in terms of the conceptual roles of all the sentences in which it appears. There may be simple characterizations of the conceptual roles of some expressions. For example, the role of negation is specified by the probability laws involving negation.

McGinn claims that two component theories, containing a truth theory and a Fieldian conceptual role theory, can deal with the problems we discussed in the first section (McGinn, 1982, pp. 234–7, 247). Consider Arabella and Twin Arabella. Their languages, English and Twin English, are syntactically identical, and since the twins are physically type-identical, the conceptual roles of their expressions are isomorphic. Each one's sentence 'water is wet' has the same (or isomorphic) conceptual role(s), so their mental states are identical. However, their sentences differ in their truth conditions, since *the fact* that makes Arabella's sentence true is H_2O's being wet, while *the fact* that makes Twin Arabella's sentence true is XYZ's being wet. A similar remark can be made concerning indexicals. The sentence 'I am thirty years old' has the same conceptual role for Arabella and Twin Arabella, but tokens of the two differ in their truth conditions. Since it is the conceptual role of an expression (in the language of thought) that determines its role in the production of behavior, Arabella and Twin Arabella will behave identically when each believes what she would express by saying 'I am thirty years old'. But, since conceptual role does not determine truth conditions, the truth values of their beliefs may differ.[2] The dual-component view seems also able to account for sentences which apparently have the same truth conditions but differ in meaning. Thus 'Cicero is bald' and 'Tully is bald' are supposed to have the same truth condition, but a given speaker's probability assignment might contain an S such that P ('Tully is bald'/S) $\neq$ P ('Cicero is bald'/S).

There is a number of features in the Field–McGinn characterization of conceptual role that are worth noting.

(1) McGinn's account differs from the kinds of accounts suggested by Sellars, Harman, and Block, in which conceptual role is characterized in terms of the *causal* relations that hold among representational mental states, perceptions, and behaviors. An individual's probability assignment does not specify causal relations, but rather his commitments concerning rational change of belief. Only if the proba-

---

[2] McGinn envisions a probability assignment defined over a language which contains indexical sentences. But there are problems with interpreting probabilities of an indexical sentence. One difficulty is accounting for changes in the probability of, e.g., 'The meeting begins now', as time passes. We also note that a probability assignment does not capture the indexicality of indexical sentences, i.e. the way truth value depends on context.

bility assignment is reflected in causal relations among belief states, and the like will conceptual role be capable of functioning in psychological explanations.

(2) Field's (1978) account of conceptual role obviously involves a great deal of idealization. Gilbert Harman has argued that it is unrealistic to suppose that an individual reasons in terms of probabilities, since this would require keeping track of an enormous amount of information and would take an enormous number of computations. Perhaps this objection can be met (Jeffrey, 1983), but a more difficult problem is presented by the evidence which shows that our beliefs do not conform to probability theory. For example, people will often assign a higher probability to a conjunction than to either of its conjuncts, and they will not typically change beliefs in accordance with conditionalization. So McGinn's theory of conceptual role might not apply to human thought.

(3) Conceptual role is a holistic notion. In characterizing the conceptual role of a sentence, one must simultaneously characterize the conceptual roles of all other sentences. Any change in the probability function—even just extending it to a new vocabulary—results in a change in conceptual role for every sentence. Because of this, two people will seldom assign the same conceptual role to syntactically identical expressions. Field explicitly offers conceptual role only as an account of intraindividual meaning. He does not think that it makes sense to compare different individuals' conceptual roles (Field, 1977). However, McGinn apparently thinks that it is meaningful to make interpersonal comparisons between conceptual roles. For example, he speaks of Arabella and her twin's mental representations as having the same conceptual roles.

(4) Field's characterization of conceptual role is solipsistic since it is entirely in terms of ingredients within the mind of the individual. It is this feature that suggests to McGinn that conceptual role can provide an account of the aspects of meaning that meets Fodor's methodological solipsism constraint. On some other versions, for instance Harman's, the characterization of conceptual role also includes relations among sentences, environmental features, and behaviour. So Harman's (1986) account of conceptual role, as he insists, is non-solipsistic. We can imagine a theory between Field's and Harman's (along this dimension), which includes relations to sensory inputs and behavioral outputs in the characterization of conceptual role. As long as the inputs and outputs are described in ways that do not entail the existence of anything other than the thinker's body, the solipsistic nature of the account is preserved. The difference will be important when we come to evaluate the adequacy of conceptual role theories as semantic theories for languages of thought.

(5) It should be clear that Putnam's stereotype and Kaplan's character are quite different from conceptual role (and from each other). The stereotype of 'water' is the information which a typically competent speaker of English

associates with 'water'—say, that water is liquid, necessary for life, fills the oceans, and so on. Stereotype differs from conceptual role in a number of ways:

(a) Stereotype characterizes cognitive significance, since it specifies the information associated with a term. It is not obvious that the conceptual role of a term or sentence associates with it any information (see fourth section).

(b) It is not clear that stereotype is 'in the head' in the way conceptual role is. The expressions used to characterize the information contained in stereotype are themselves subject to Twin Earth arguments, and this seems to show that stereotype itself is not solipsistic.

(c) As we have pointed out, conceptual role is holistic. Stereotype does not appear to be holistic.

There are also important differences between conceptual role and character:

(a) Two people can associate the same character with 'I am hungry', even though the sentence has different conceptual roles for each, since the two may differ in their overall probability assignments. So, character can be used to explain the sense in which two people who assert 'I am hungry' share the same belief, while conceptual role cannot.

(b) Character determines truth conditions relative to context, but there is no systematic relation between conceptual role, context, and truth conditions. At least none is built into McGinn's account.

No matter how McGinn might fill in the details of his dual-component view, the general picture is clear. On his view, the appropriate form for a semantic theory for a language is a conjunction of *two* theories: one characterizes internal mental features of meaning, the other characterizes relations between language and the world. In opposition to the dual-component view are semantic theories which provide a *unified* treatment of the mind aspects and world aspects of meaning. Frege's and Davidson's theories are examples. McGinn, of course, thinks that such unified accounts are misguided. He says: 'But it seems that nothing of critical importance would be lost, and some philosophical clarity gained, if we were to replace in our theory of meaning, the ordinary undifferentiated notion of content by the separate and distinct components exhibited by the conjunctive paraphrase' (McGinn, 1982, p. 229). We will argue in the fourth section that, contrary to McGinn's claim, something of critical importance is missed by bifurcating the theory of meaning in the way McGinn proposes. Dual-component theories cannot be used as theories of interpretation, and for this reason they fail to provide adequate accounts of communication. We will present this argument in the final section. First, we want to show how Davidsonian truth theories can be used as theories of interpretation.

Truth conditional theories of communication and understanding

According to DATs, a theory of truth for a language is incomplete qua theory of meaning because it fails to provide an account of the mental aspects of meaning: language understanding and cognitive significance. The conceptual role component is supposed to do that job. This view of the place of a truth theory in an account of meaning is clearly at variance with Donald Davidson's. Davidson sees a truth theory as being capable of providing both an account of language understanding and an account of the relations between language and reality. In this section we will show how it is that knowledge of truth conditions, specified in a familiar Davidsonian way, can play a central role in understanding and communication. Our argument is a bit different from the arguments found in Davidson, but we clearly take our cue from his writings.

It is almost a truism in the philosophy of language (or so it was until recently) that to understand a sentence is to know its truth conditions. But, if it is a truism, it is an obscure one. We will try to show what truth it contains. Once again, consider Arabella. She utters the words 'Es schneit' within earshot of Barbarella and Cinderella.[3] Barbarella understands German while Cinderella does not. This makes a difference. Barbarella acquires the beliefs that it is snowing and that Arabella believes that it is snowing, and perhaps some other beliefs as well. Cinderella does not acquire these beliefs; Arabella's utterances are so many sounds to her. Even if she recognizes them as forming assertion, something she is able to do without understanding German, she may acquire only the belief that Arabella's utterance 'Es schneit', whatever it may mean, is true, and perhaps also the belief that Arabella holds her utterance to be true. Still, she does not know what Arabella expresses or believes.

We have argued (Lepore, 1982b; Loewer, 1982; Lepore & Loewer, 1983) that *a theory of meaning for a language L should include information such that someone who possesses this information is, given his other cognitive capacities, able to understand L.* Understanding a language involves many complex abilities, such as to respond appropriately to assertions, orders, questions, and so forth. We will focus on one central ability: the ability to acquire justifiable beliefs about the world and about each other's beliefs through our linguistic practices. Since Cinderella, who does not understand German, can come to know that Arabella's utterance 'Es schneit' is true, we can ask what additional information would enable her to acquire, justifiably, the beliefs which Barbarella acquires.

A plausible (indeed, we think an inevitable) answer to our question is that, if Cinderella knew that 'Es schneit' is true (in German) if and only if it is snowing, she would be in a position to acquire the target beliefs.

---

[3] We switch languages here from English to German only to emphasize the informational value of T sentences.

The reasoning for justifying these beliefs could go as follows:

Paradigm I:
(1) Arabella's utterance 'Es schneit' is true.
(2) 'Es schneit' is true iff it is snowing.
So,
(3) It is snowing.
Paradigm II:
(4) Arabella believes 'Es schneit' is true.
(5) Arabella believes 'Es schneit' is true iff it is snowing.
So,
(6) Arabella believes that it is snowing.

Elsewhere we have argued that such reasoning gives substance to the claim that to understand a sentence is to know its truth conditions (Lepore, 1982b, 1983; Lepore & Loewer, 1981, 1983). However, this claim requires some qualification. We are not saying that a person's understanding of German involves his going through the above inferences, or even that every person who understands German explicitly knows the truth conditions of German sentences. Our claim is that truth conditions explicitly state information which can be used (normally together with other information) to interpret utterances. In this way, a specification of truth conditions for a language can provide an illuminating characterization of language understanding and communication.

The view that a theory of truth for L can serve as a theory of meaning for L is most prominently associated with Donald Davidson. Our two paradigms exhibit exactly how a theory of truth can play the role of a theory of meaning. But, of course, not just any theory which entails for each indicative sentence S of L a theorem of the form 'S is true iff p' can serve as a theory of interpretation for L. According to Davidson, a theory of interpretation for an individual's language should assign truth conditions to her utterances in a way that results in an attribution of beliefs and preferences to her which are *reasonable* given her situation and behavior. Exactly what counts as reasonable will depend on our theories of belief and desire acquisition and on our theories of behaviour (Davidson, 1973b, 1974). We would add that a theory of truth together with a theory of intepretation should yield theorems which can be employed in our paradigms. While we have not developed the adequacy conditions on Davidsonian truth theories in detail, what we have said is sufficient to distinguish among truth theories for L. For example, they may serve to eliminate theories which entail '"Schnee ist weiss" is true if grass is green.' If we use a theory which contains this theorem, then we might infer, from Arabella uttering 'Schnee ist weiss', that grass is green and that she believes that grass is green. But Arabella might not have this belief. In any case, we would not be justified in believing that she has this belief (or that grass is green) on the basis of the truth of her utterance. It is interesting to see that our constraints also distinguish the truth conditions '"Water is wet" is true iff water is wet' from '"Water is wet iff H_2O is wet'. The latter, but not the former, license an inference from Arabella, who believes that her utterance 'Water is wet' is true, to 'Arabella believes that H_2O is wet'. We can imagine circumstances in which this would lead to error—that is, cirumstances in which Arabella fails to believe that water is H_2O.

Readers familiar with discussions of Davidson's accounts of language will notice that we have emphasized the importance of knowledge of truth conditions, while we said little concerning the nature of the theory which implies instances of (T). Some writers, for example Harman, claim that, whatever a truth theory has to say about meaning, it is contained in the recursion clauses of the theory which show how truth conditions of complex sentences depend on semantic features of their component expressions (Harman, 1974; also Fodor, 1975). Harman argues that such a theory, at best, characterizes the meanings of logical constants—'and', 'or', and so on—by characterizing their conceptual roles, but that theory does nothing to specify the meanings of other expressions. It should be clear that we disagree (see Lepore & Loewer, 1981). Truth conditions do specify meanings in that they enable someone who knows the truth conditions of sentences to interpret the speech of another. Of course, the theory is important as well, but not because it characterizes the conceptual roles of the logical connectives. (It is not clear that it does: Lepore, 1982a.) Having a truth theory for L is important because it provides a specification of truth conditions for all the (infinitely many) sentences of L in a way that does not presuppose an understanding of L.

We claim that truth theories for natural languages which are theories of interpretation address both aspects which concern DATs. This is clearly seen from our paradigm inference patterns. Someone who knows the truth conditions of the sentences of a language and knows that this is common knowledge among speakers of that language is in a position to draw conclusions both about the world and about what other speakers have in mind. On the one hand, truth conditions relate sentences to the world. They specify what must hold for a sentence to be true. On the other hand, they specify what is known by someone who understands a language. Whether they can deal with the specific problems that motivated DATs remains to be seen. We will discuss this matter in the last section.

What's wrong with DAT?

In this section we will compare DATs, focusing mainly on McGinn's version, with Davidsonian truth theories. We will argue for three claims:

(1) McGinn's account of 'S means that p' involves necessary conditions for sameness of meaning which are much too restrictive and render it incapable of providing an account of communication.

(2) Neither component of a DAT, nor the two together, is a theory of interpretation.

(3) The conceptual role component of a DAT does not supply the sort of semantics for the language of thought that is required by (Fodor's version of) cognitive science.

McGinn's (1982, p. 229) analysis of meaning is that:

> for S to mean that p is for S to be true iff Q, for some 'Q' to have the same truth conditions as 'p', and for S ϕ to have some cognitive role ϕ such that 'p' also has cognitive role ϕ.

McGinn's analysis specifies that S and S★ have the same meaning only if they have identical conceptual roles. Field explicitly claims that sameness of conceptual role is a necessary condition for intrapersonal synonymy, but also explicitly denies its usefulness in characterizing interpersonal synonymy (Field, 1977). McGinn is unclear on this point, but if his analysis of 'S means that p' is to be used by an interpreter to specify the meanings of a speaker's sentences, then it requires that the conceptual role of the interpreter's sentence 'p' and the conceptual role of S be identical. However, only in the kinds of thought experiments found in science fiction are the conceptual roles of Arabella's and Barbarella's sentences the same. As long as there is the slightest difference between Arabella's and Barbarella's probability assignments, no sentence will have the same conceptual role for both. If Arabella assigns a probability of 1 to 'it is raining' while Barbarella assigns it a probability of 0.2, then, on McGinn's account, their sentences have different meanings. Barbarella would be mistaken if she said that Arabella's sentence 'it is raining' means that it is raining. It is difficult to see how sense can be made of communication on this account of sameness of meaning. If Arabella and Barbarella assign different meanings to 'it is raining' (because they assign different probabilities), then there is nothing in common between them, to be communicated. If they assign the same meaning (and hence have the same probability assignment), then there is no need for communication.

McGinn's account runs into similar difficulties as an account of 'means that' for internal representations. The usual view of advocates of Mentalese accounts of belief (such as Fodor) is that 'Arabella believes that p' is true if and only if she bears a certain relation R to a representation S which means that p. Since it is improbable that any of Barbarella's sentences has the same conceptual role as S, Barbarella's claims concerning what Arabella believes are bound to be incorrect on McGinn's view.

The heart of the problem is that, for S to mean that p, S need not have precisely the same conceptual role as 'p', but rather a conceptual role appropriately similar to 'p'. Until the DAT theorist has specified when the conceptual roles of S and 'p' are sufficiently similar for S and 'p' to count as having the same meaning (when they have the same truth condition), he has not adequately characterized 'S means that p'. But this seems to be a hopeless task as long as conceptual role is characterized purely formally—for instance in terms of probability relations.

Even if McGinn could overcome the difficulties just discussed, we will now show that the two components of McGinn's DAT, either taken separately or together, fail to comprise a theory of interpretation. The adequacy conditions to be met by a DAT truth theory differ significantly from those to be met by a Davidsonian truth theory. McGinn views the truth theory as assigning 'facts' or 'states of affairs' to be indicative sentences. This leads him to see a theory which issues in, say, ' "water is wet" is true iff H_2O is wet' as adequate since, according to him, 'water is wet' and 'H_2O is wet' express the same fact. In fact, he says that any statements that are necessarily equivalent (i.e. substitutable in any non-intensional context) have the same truth conditions. Field's conception of the truth-theory component is a bit different. He imagines the theory issuing in theorems like: 'water is wet' is true iff the stuff denoted by 'water' has the property denoted by 'is wet'. He hopes for a physicalistic theory of the denotation relation. It is clear that McGinn and Field see a truth theory as explicating the relation between language and the world. McGinn thinks that sentences are assigned facts. Field sees expressions as denoting bits of reality.

The first point to notice is that neither McGinn's nor Field's truth theories will serve as theories of interpretation (except in very unusual circumstances). We will discuss McGinn's account first. Suppose that Arabella utters 'water is wet'. If Barbarella knows that 'water is wet' is true if and only if H_2O is wet, then she will correctly conclude (following our first paradigm) that H_2O is wet. However, following our second paradigm, she may also conclude that Arabella believes that H_2O is wet, and this might very well be a mistake. The trouble is that, even though 'water is wet' is true if and only if H_2O is wet, this truth condition does not express the belief Arabella intends to communicate. Field's truth theory is even less adequate as a theory of understanding. If Barbarella knows that 'water is wet' is true if and only if the stuff denoted by 'water' has the property denoted by 'is wet', then she will be able to conclude from Arabella's utterance that the stuff denoted by 'water' has the property denoted by 'is wet'. But there is a large gap between this and the conclusion that water is wet. It is certainly the latter, and not the former, that is communicated by Arabella's utterance.

The previous objections show that the truth-theory component of a DAT does not provide a full specification of the information which a competent English speaker has, and brings to bear, in interpreting the utterances of others. Thus, it does not express the information known by a competent speaker. But, of course, Field and McGinn do not advance these theories as theories of understanding. Since the conceptual role component is supposed to characterize the mental aspects of meaning, perhaps it specifies information sufficient to interpret a speaker's utterances. So we will now examine the conceptual role theory, to see whether it will enable one to interpret another's language.[4]

We will first discuss McGinn's and Field's proposal concerning the representation of conceptual role, and then the less precise characterizations given by Sellars and

---

[4] It is obvious that the truth-theory component of a DAT does not assign content to sentences of Mentalese in a way that respects the formality condition. DAT assignments of truth conditions do not respect our ordinary individuation criteria for belief attributions. A representation R might be assigned either the truth condition that water is wet or, equivalently from the perspective of a DAT, that H_2O is wet. While there may be a 'transparent' sense of belief for which the belief that water is wet is the same as the belief that H_2O is wet, it is clear that this assignment of belief contents is inappropriate for CTM.

Harman. We will suppose, once again, that Arabella utters 'Es schneit'. We will also suppose that Cinderella knows the conceptual roles of all of Arabella's sentences; that is, she knows the conditional probability function $P(A/B)$ defined on all the sentences of Arabella's language. Is this information sufficient to enable her to infer justifiably either that it is snowing or that Arabella believes that it is snowing? The answer is no! Suppose that Cinderella starts by knowing that Arabella believes 'Es schneit' is true (or perhaps that she assigns a probability close to 1 to 'Es schneit'). How can the conceptual role theory support Cinderella's inference to the belief that it is snowing? It may support Barbarella's concluding that Arabella assigns a high probability to 'Es ist kalt', a low probability to 'Das Wetter ist schoen', and so on. She might even be able to predict how Arabella would behave, what sounds she will utter next, and so on; but she will not know what Arabella means. In particular, she will not be in any position to infer that it is snowing, or that Arabella believes that it is snowing. Unless Barbarella knows that 'Es schneit' is true if and only if it is snowing, Arabella's assigning a probability of 1 to 'Es schneit' does not count as evidence that it is snowing. Nor does it count as evidence for Arabella believing that it is snowing. Knowledge of Arabella's probability assignments is not sufficient for interpreting her utterances, at least not if that understanding includes the capacity to infer that it is snowing, that Arabella believes that it is snowing, and so on.

One might think that the problem with Field's conceptual role semantics (as a characterization of knowledge sufficient for language understanding) is that it fails to characterize relations between expressions and the world. In contrast, Sellars and Harman sketch a conceptual role theory which includes causal relations between sentences and the world. Harman says:

There is no suggestion that content depends only on functional relations among thoughts and concepts, such as the role a particular concept plays in inference. Of primary importance are functional relations to the external world in connection with perception, on the one hand, and action on the other. (1982, p. 247)

Harman does not specify the structure of a theory which characterizes conceptual role non-solipsistically. Lycan (1984, Ch. 10), in the course of criticizing Harman's proposal, suggests that such a theory would imply statements like:

(A) The usual cause of Arabella's uttering 'Es schneit' is her being in belief state K, and the usual cause of her being in belief state K is it's snowing in the vicinity.

If Barbarella knew this, then she would be in a position to infer that it's snowing from Arabella's having uttered 'Es schneit'. This would be a case of inference to the best explanation, rather than an instance of our first paradigm. Truth seems to play no special role. As Lycan is well aware, there is a number of problems with this proposal. It is not plausible for most sentences, especially theoretical ones. The usual cause (if there is such a thing) of my uttering 'there are positively charged electrons' is not my being in a belief state which is itself caused by there being positively charged electons.

The latter state of affairs may figure in the causal history of my utterance; but it is only one among many causes, and not the 'usual cause'. If we put this objection aside, (A) could be used to draw conclusions about the world from Arabella's utterances. But it does not enable one to draw appropriate conclusions about what Arabella believes when she utters, say, 'water is wet'. The problem is that, if water's being wet is the cause of her utterance, so is H_2O's being wet. In fact, Lycan's proposal adds nothing more to the theory of interpretation than the truth theory component provides.

We can imagine McGinn replying to our discussion with the suggestion that, although neither component of a DAT is, by itself, a theory of interpretation, their combination is. One might be encouraged to think this since, according to McGinn, the conjunction of the two theories and his definition of meaning (M) yield theorems of the form 'S means that p' for each sentence S of L. Since 'S means that p' entails 'S is true iff p', we could employ these theorems in our two paradigms. But a closer look at McGinn's analysis of 'S means that p' reveals that the two component theories do not entail statements of the form 'S means that p'. Recall McGinn's definition:

> (M) S means that p iff S is true iff Q, for some 'Q' having the same truth conditions as 'p', and for S to have some cognitive role Q such that 'p' also has cognitive role Q. (1982, p. 229)

The truth-theory component may entail, for instance, that 'Wasser ist naß' is true if and only if H_2O is wet. The conceptual role component may entail that 'Wasser ist naß' has the same conceptual role as 'water is wet' (and a different conceptual role from 'H_2O is wet'). But we can put the two components together to explain that 'Wasser ist naß means that water is wet only if we know that 'Wasser ist naß, means that water is wet. Of course, we know this if we know English. But McGinn cannot rely on this, since a theory of interpretation is supposed to characterize those features knowledge of which would be sufficient for understanding a language, and it should do this in a way that does not already presuppose understanding the language in question. The difficulty here is the same as the one we encountered in considering the conceptual role component by itself. Knowing that S has the same conceptual role as 'p' is not sufficient for understanding S (Twin Earth problems aside), unless one knows what 'p' means. Contrast this with knowing that S is true if and only if p. Someone might have this knowledge (the knowledge expressed by 'S is true iff p') without knowing what 'p' means. We want the theory of interpretation to state information sufficient for interpretation, which itself does not depend on understanding the language in which it is formulated. It is this that (M) fails to do. Our conclusion is that the Field–McGinn conceptual role theory is inadequate as a theory of interpretation.

What is the relation between McGinn's two-component theory and a Davidsonian truth theory? According to McGinn (1982, p. 240),

the Davidsonian perspective, while not actually being incorrect—for it is, after all, tacitly a dual component conception—is apt to deceive us about the theoretical resources we need in a full adequate theory of meaning.

McGinn thinks that a Davidsonian theory is 'tacitly a dual component conception', since in a Davidsonian T-sentence of the form 'S is true iff p', p not only specifies a truth condition of S, but is also supposed to be a translation of S. If translation is understood as requiring identity (or, rather, isomorphism) of conceptual role, then it seems to McGinn that a Davidsonian theory is simply a misleading formulation of a two-component theory.

McGinn is correct in thinking that a Davidsonian theory addresses two aspects of meaning. It accounts for language–world relations by characterizing truth and reference, and it accounts for mental aspects by characterizing knowledge which is sufficient for interpreting a language. But we strongly disagree with McGinn's claim that Davidson's theory is tacitly a dual-component view presented in a misleading manner. There are significant differences between a Davidsonian theory and McGinn's DAT. First, a Davidsonian theory does not require, for the adequacy of its T-sentence theorems, that S and 'p' have the same conceptual role. We have already discussed the implausibility of this requirement. Davidson's theory is subject to the much looser constraint that it leads to reasonable attributions of beliefs and other propositional attitudes to speakers. Second, as we have just argued, McGinn's DAT is not a theory of interpretation. Knowledge of it does not enable one to determine what Arabella believes on the basis of her sincere utterances. This is the heart of our objection to McGinn's and other two-component theories. Whatever insight, if any, may be gained from decomposing meaning into two separate features, the resulting characterizations are not suitable as a theory of interpretation. Knowledge sufficient for understanding a language cannot be extracted and separated from knowledge of reference and truth, since to understand a sentence is to know what it says about the world.

We have yet to consider the question of whether DATs provide adequate accounts of meaning for mental representations. It is prima facie plausible that DATs are just what is needed for Mentalese. Recall that Fodor argues that cognitive science requires a methodologically solipsistic notion of content in its explanations of behavior. It is not unreasonable to look to the conceptual role component of a DAT to characterize this notion of content, while the truth-theory component characterizes a broader notion of content involving relations between mental representations and the world. Since Mentalese is not a language in which anyone communicates, our arguments that DATs do not provide theories of interpretation may appear to be irrelevant. In fact, some authors explicitly endorse dual-aspect accounts for Mentalese, but not for public languages (Lycan, 1985).

In the course of our discussion of Fodor in the first section, we saw that Fodor thinks that cognitive science requires a solipsistic notion of content. That is, a characterization of content which conforms to the formality condition: no difference in content without a difference in formal properties. The Field–McGinn characterization of conceptual role is solipsistic. But, we will argue, it does not yield a specification of content that is methodologically solipsistic, for the simple reason that *it yields no specification of content at all*. We have, in effect, already established this in our discussion of DATs as theories of interpretation. The problem is that a complete specification of the conceptual roles of the sentences S of Arabella's language of thought does not enable us to fill in the blanks in 'If Arabella bears R★ to S, then she believes that——'. But this is what Fodor requires of a characterization of the contents of mental representations.

It might be thought that the dispute between us and someone who holds that conceptual role provides a solipsistic notion of content is merely a *semantic* quibble about what is to count as content. But it is not, since the operative notion of content is intended (by Fodor, who formulated this problem) to play a particular role in psychological explanations. Consider the following 'psychological' explanation. Arabella jumped because she believed that by jumping she would cause it to rain, and she wanted it to rain. The phrases following 'believed that' and 'wanted' express contents. They explain Arabella's action by citing causes (her beliefs and desires) which 'rationalize' it. If we describe the causes of her actions without citing the contents of the propositional attitudes, then the resulting explanation no longer has its 'rationalizing' force. An explanation of Arabella's jumping which employs conceptual role would go somehow like this. Arabella bears B★ to a mental representation which has conceptual role Q and W★ to a mental representation which has conceptual role V; so, she jumped. Presumably the conceptual roles would be characterized in such a way that jumping normally follows upon bearing these relations to those representations. It is clear that the explanatory force of this explanation, whatever its value, does not rationalize Arabella's behavior. Only an explanation which appeals to content can do that.

Davidsonian truth theories solve the problems

We have shown that McGinn's two-component definition of meaning is inadequate, that his DAT is not suitable for use as a theory of interpretation, and that it fails to yield a solipsistic notion of content for Mentalese. Also, we began to show how a Davidsonian truth theory, by providing a unified account of understanding and reference, is suitable for use as a theory of interpretation. Still, we need to inquire into the extent to which Davidsonian theories can cope with the problems which motivated the DAT proposal. So we will now consider whether a Davidsonian truth theory can be used to interpret sentences containing proper names, indexicals, and terms for natural kinds. We will conclude with a few remarks on the prospect of obtaining a methodologically solipsistic theory of content.

What T sentences will a Davidsonian truth theory entail for a language with proper names? It is clear that, in contrast to the truth-theory component of a DAT, a

Davidsonian theory must be able to assign different truth conditions to, say, 'Cicero is bald' and 'Tully is bald'. DAT truth theories maintain that these sentences have the same truth conditions, since it is metaphysically necessary that they are equivalent (assuming that names are rigid). If one thinks, as McGinn does, that a truth theory assigns possible states of affairs or facts to indicative sentences, then we can see why DAT truth theories assign the same truth conditions to the two sentences. But it is not necessary to think of truth theories in this way. It is clear that Davidson rejects the reification of truth conditions as states of affairs (Davidson, 1969).

There is no reason why a Davidsonian truth theory cannot contain (1) and (2) without containing (3) and (4) (McDowell, 1980):

(1) 'Cicero is bald' is true iff Cicero is bald.
(2) 'Tully is bald' is true iff Tully is bald.
(3) 'Cicero is bald' is true iff Tully is bald.
(4) 'Tully is bald' is true iff Cicero is bald.

Arabella might believe (1) and (2) without believing (3) and (4). In that case it would be a mistake to include (3) and (4) in a theory intended to be employed in interpreting her utterances. If (3) and (4) are included in a truth theory for Arabella's language, then one would be licensed to infer (via paradigm II) that Arabella believes that Tully is bald from the fact that she utters 'Cicero is bald'. Suppose that a truth theory T contains (1) and (2), but not (3) and (4), and that Arabella assents to 'Cicero is bald' and dissents from 'Tully is bald'. We can use T to conclude that Arabella believes that Cicero is bald and believes that Tully is not bald. But there is no contradiction forthcoming unless we have some principle in the theory that permits the substitution of co-referential names in belief contexts. That it leads to attributing contradictory beliefs to Arabella is sufficient reason for rejecting the principle.

Our account of the truth conditions of sentences containing names is compatible with the view that names rigidly designate. If names are rigid designators, then true identity statements composed of names are necessarily true. Suppose that we add to T the 'axiom'—It is necessary that (Cicero = Tully). We can also suppose that Arabella knows that, if Tully = Cicero, then it is necessary that (Tully = Cicero). Of course, we will not add the axiom that Arabella knows that it is necessary that (Cicero = Tully), since that is false. Even with these additions, we cannot derive from T—together with the fact that Arabella assents to 'Cicero is bald' and dissents from 'Tully is bald'—that she believes that Cicero is bald and believes that Cicero is not bald. Of course, if one thinks that co-referential rigid designators are substitutable *salva veritatae* in belief contexts, then the consequence will be to saddle Arabella with contradictory beliefs. As we mentioned earlier, that seems to us to be a reason for rejecting substitutivity.

A defender of DATs is likely to respond to our proposal in the following fashion. A Davidsonian truth theory is supposed to be an empirical theory, but what evidence could distinguish between a theory which contains (1) and (2) and a theory which contains (3) and (4)? Our reply, of course, is that the two theories are empirically discriminated in their applications in paradigm II. Assuming that Arabella might believe that Tully is bald without believing that Cicero is bald (and that we can obtain evidence for this), we can distinguish between the two theories. What evidence is relevant to whether Arabella believes that Tully is bald or believes that Tully is not bald? The dual-aspect theorist's answer is that conceptual role is relevant. If Arabella bears B⋆ to an internal representation 'Tully is bald' which has a conceptual role similar to the conceptual role of my sentence 'Tully is bald', then she believes that Tully is bald. But, as we argued previously, identity, or similarity, of conceptual role is too strong a requirement to make us co-believers. Conceptual role seems especially irrelevant when it comes to translating proper names. What does seem relevant is the history of the acquisition of a name. Suppose for example that I can trace Arabella's current use of 'Cicero' and my current use of 'Cicero' back to a common source; say, we both acquired the name upon hearing it spoken in a history class. Then that would count in favour of my attributing to her the belief that Cicero is bald upon hearing her utter 'Cicero is bald', even if the conceptual role of her sentence is quite different from mine. Of course, considerations governing the interpretation of names are quite complex. Our point is that whatever they are, (a) conceptual role is not of great importance; and (b) a Davidsonian truth theorist can (indeed should) avail himself of these considerations in fashioning a truth theory.

Indexicals provide another motivation for DATs. Arabella's and Twin Arabella's understanding of 'I am thirty years old' is the same, but the truth conditions of their utterances of the sentence differ. DATs say that the conceptual roles of the sentences are the same (or are isomorphic), while the truth conditions of their utterances differ. As we observed in the second section, a conceptual role theory does not seem to provide an adequate account of our understanding of indexicals. One's understanding of, for example, 'I am now in California' involves knowing how to infer what was said from knowledge of the context in which the sentence was uttered. It is not clear that this can be represented within Field's framework for conceptual role. Be that as it may, we now want to show how truth theories can be modified to interpret sentences with indexicals.

A number of authors have proposed ways of extending truth theories to languages with indexicals (See, for example, Taylor, 1980). However, no one seems to have addressed the problem of showing how a truth theory for languages with indexicals can be employed to interpret the utterances of a speaker of the language. It may be thought that a truth-conditional account will not work for indexicals, because 'Arabella is tall' and 'I am tall' uttered by Arabella have the same truth condition—namely that Arabella is tall. But clearly Arabella's understanding of the two sentences might be different, since she might believe what one expressed and not what the other expressed. DATs deal with this by claiming that there is something other than truth conditions involved in understanding. What we are going to do is to show how a truth-conditional account of indexicals can yield an account of understanding according to which Arabella and Barbarella have the same understanding of, say, 'I am in

California', and yet different information can be communicated by their utterances of the sentence. Our idea builds on Davidson's suggestion that the truth predicate applies to a sentence at a time, for an utterer, at a place (with perhaps further relativizations required).

On our view, indexical sentences possess *general truth conditions*. For example, understanding 'I am now reading Russell' includes knowing that the following is common knowledge among speakers of English:

(C) $(x)(t)$ ('I am now reading Russell' is true for x at t iff x is reading Russell at t).

Interpreting a language with indexicals involves introducing a relativized truth predicate. With 'I' and 'now' as the only indexicals in this sentence, the relativized truth predicate is 'true for x at t', where 'x' ranges over utterers and 't' ranges over times. How can knowledge that (C) enable someone to interpret utterances? Suppose that Arabella says to Barbarella: 'I am now reading Russell'. Then Barbarella may be able to reason as follows:

(1) Arabella uttered 'I am now reading Russell' at noon.

So,

(2) 'I am now reading Russell' is true for Arabella at noon.

(C) $(x)(t)$ ('I am now reading Russell' is true for x at t iff x is reading Russell at t).

So,

(3) Arabella is reading Russell at noon.

Barbarella is justified, in part, in believing that Arabella is reading Russell at noon on the basis of her hearing Arabella's utterance, because she knows that (C). But notice that she *also* makes use of information about the context of utterance. In particular, she uses her belief that Arabella is the utterer and that she made her remark at noon. If Barbarella failed to believe this, she would not be in a position to conclude (3). Exactly what she can conclude depends on what she believes. If she believes that the utterer is the tallest woman in the room and that the time is the same as when Cinderella laughed, then Barbarella could employ (C) to learn from Arabella's remark that the tallest woman in the room was reading Russell at the same time when Cinderella laughed.

Of course, there are ways that any competent interpreter has of identifying an utterer and the time of an utterance. For example, if Barbarella hears someone utter 'I am now reading Russell', even if she knew nothing else about the utterer, she would know that the utterer is the utterer of that utterance. She could then use (C) to conclude that the utterer of that utterance of 'I am now reading Russell' is reading Russell.

We can also explain how Barbarella can employ her knowledge that (C) is common knowledge among speakers of English in order to learn something about what *Arabella* believes when she utters 'I am now reading Russell'. But here the inferences are a bit more delicate.

Considering the following reasoning:

(1) Arabella utters 'I am now reading Russell' at noon.
(4) Arabella believes that 'I am now reading Russell' is true for Arabella at noon.
(5) Arabella believes that $(x)(t)$ ('I am reading Russell' is true for x at t iff x is reading Russell at t).

So,

(6) Arabella believes that Arabella is reading Russell at noon.

In this bit of reasoning Barbarella employs information concerning Arabella's beliefs about who she is and about when her utterance occurs. To conclude (6), she uses the information that *Arabella* believes that *Arabella* produced the utterance and that *Arabella* believes that the reading of Russell in question occured at noon. If Arabella did not know that she is Arabella or did not know that the time of the utterance is noon, then Barbarella would be mistaken in concluding that (6).

If Barbarella does not know any singular term (or a translation thereof) that she believes Arabella believes refers to herself, then she will be unable to employ the knowledge embodied in the general truth condition (C) to arrive at conclusions about when Arabella believes herself to be reading Russell at noon. Again, there are some ways that any competent speaker of English has of identifying herself and the time of her utterance. Arabella will always know (if she understands English) that 'I' refers to herself. Of course, Barbarella cannot use 'I' to refer to Arabella, but she can express what Arabella believes by ascribing to her the belief that the utterer of that utterance believes of *herself* that she is reading Russell, from which she can conclude that Arabella believes that she (herself) is reading Russell.

Knowledge that (C), together with other information (some available to every competent speaker, some not), enables an interpreter to learn both about the utterer's beliefs and about the world beyond the utterer. Suppose that, as in Perry's story, Hume and Heimson each utter 'I am a Scottish philosopher' (Perry, 1979). Each one believes himself to be Hume. If we know all this, we can employ the general truth condition—(x) ('I am a Scottish philosopher' is true for x iff x is a Scottish philosopher)—to conclude from Hume's utterance that Hume is a Scottish philosopher and that he believes himself to be a Scottish philosopher. Since we know that Heimson is deluded, we would not conclude from his utterance that he is a Scottish philosopher. But we can conclude, following the second inference pattern above, that Heimson believes himself to be a Scottish philosopher.

We have indicated how general truth conditions can play a role in a theory of interpretation. To provide an adequate account, we would at least have to show how to construct a truth theory for a language with indexicals, which has correct general truth conditions as theorems; and we would need to develop a logic in which the inferences to conclusions about the world and the utterer's beliefs can be represented. We leave these tasks to a future paper.

Natural-kind terms presented another problem which motivated the development of DATs. Can a truth theory of interpretation be constructed for languages containing natural-kind terms? Imagine an English-speaking radical interpreter who arrives on Twin Earth. He notes the patterns of sentences held to be true (and the degrees to which they are held to be true) and begins to devise a theory of interpretation. Suppose that he is unaware that the stuff they call 'water' is composed of XYZ molecules. Then he is likely to employ the homophonic truth theory, which contains the clause ''Water is wet' is true (as uttered by a Twin Earther) iff water is wet'. When Twin Arabella utters 'This is water', the interpreter will conclude that this is water and that Twin Arabella believes that this is water, just as he would conclude if he were interpreting Arabella on Earth. But he will be mistaken. This is not water (since it is composed of XYZ), and Twin Arabella does not believe that it is water (since she has never encountered water). However, these mistakes will pass undetected as long as the interpreter does not know the compositions of water and twin water. Once he discovers that the chemical composition of water is H_2O and the composition of twin water is XYZ, he will notice that he sometimes interprets Twin Arabella incorrectly. How might he revise his theory in the face of these incorrect interpretations?

He could try '"Water is wet" is true in Twin English iff XYZ is wet'. This change will lead him to conclude correctly that this is XYZ when he hears Twin Arabella utter 'This is water', but it might also lead him to conclude mistakenly that Twin Arabella believes that this is XYZ. This would be a mistake if Twin Arabella did not know the chemical composition of the stuff she calls 'water'.

The solution is for the interpreter to enrich his own language by adding a term 'water^t' with the stipulation that it rigidly designates XYZ, but has the same conceptual role as Arabella's word 'water'. His truth theory for Twin English should contain the theorem '"This is water" is true in Twin English iff this is water^t'.

We have shown how understanding a language which contains names, indexicals, and natural kind terms can be characterized by the knowledge of a truth theory which satisfies certain constraints. We argued that the DAT account is not successful, since the conceptual role component cannot explain how we acquire information from the utterances of others. However, it is clear that our Davidsonian account of understanding does not satisfy one of the constraints that the conceptual role component was designed to satisfy; on the Davidsonian account, meaning is not entirely in the head. Arabella and Twin Arabella may be physiologically identical, but the truth theories that describe their understanding of their languages are different, as we saw in interpreting natural-kind terms in English and Twin English. We have argued elsewhere (Lepore & Loewer, 1986) that this does not show Davidsonian truth theories to be inadequate, but rather that the information that an interpreter can use does not supervene on physiology. The view that it is possible to describe a person's understanding of a language at a level at which understanding supervenes on physiology and which is also able to account for communication is wrong.

Conclusion

Dual-aspect theories claim that there are two aspects to meaning: one which relates language to the world and another which relates language to the mind. McGinn goes on to claim that the most perspicacious theory of meaning is one which gives distinct and independent treatments of each aspect: a theory of truth and a theory of conceptual role. We agree that meaning has two aspects, but we disagree concerning the appropriate form for a theory of meaning. In this chapter we showed that McGinn's dual-component theory is incapable of serving as a theory of interpretation, and for that reason is not the most perspicacious form for a theory of meaning. McGinn confuses the evidential basis for a theory of meaning, which may include information about the causal relations between representations and the world as well as information about conceptual role, with a theory of meaning. In contrast, Davidsonian truth theories can accommodate both aspects while providing a theory of interpretation.

Acknowledgments

We would like to thank Ned Block, Jerry Fodor, Brian MacLaughlin, Richard Grandy, Gilbert Harman, Bernard Linsky, John McDowell, Colin McGinn, Hilary Putnam, and Stephen Schiffer, for comments on, and discussion of, earlier drafts of this chapter. We especially would like to thank John Biro, Donald Davidson, and William Lycan for their help. Earlier drafts of this chapter were read at the University of Michigan, Central Michigan University, Rice University, The Pacific APA (1984) the Florence Center for the Philosophy and History of Science, the University of Urbino, and the City University of New York.

References

Block, N. (1986) Advertisement for a Semantics for Psychology. *Midwest Studies in Philosophy* 10: 615–678.

Davidson, D. (1967) Truth and Meaning. *Synthese* 17: 304–323; reprinted in Davidson (2001).

Davidson, D. (1968) On Saying That. *Synthese* 19: 130–146; reprinted in Davidson (2001).

Davidson, D. (1969) True to the Facts. *Journal of Philosophy* 66: 748–764; reprinted in Davidson (2001).

Davidson, D. (1973a) In Defense of Convention T. In H. Leblanc (ed.), *Truth, Syntax and Modality*, Amsterdam: North-Holland Publishing Co., pp. 76–86, reprinted in Davidson (2001).

Davidson, D. (1973b) Radical Interpretation. *Dialectica* 27: 314–328, reprinted in Davidson (2001).

Davidson, D. (1974) Belief and the Basis of Meaning. *Synthese* 27: 309–323, reprinted in Davidson (2001).

Davidson, D. (2001). *Inquiries into Truth and Interpretation*, 2nd edn, Oxford: Oxford University Press.

Field, H. (1977) Logic, Meaning and Conceptual Role. *Journal of Philosophy* 7: 379–409.

Field, H. (1978) Mental Representation. *Erkenntnis* 13: 9–61.

Fodor, J. (1975) *The Language of Thought*. Cambridge, MA: Harvard University Press.

Fodor, J. (1980) Methodological Solipsism Considered as a Research Strategy in Cognitive Psychology. *Behavioral and Brain Sciences* 3: 63–73.

Fodor, J. (1982) Cognitive Science and the Twin-Earth Problem. *Notre Dame Journal of Formal Logic* 23: pp. 98–118.

Fodor, J. (n.d.) Narrow Content [= NC]. Unpublished manuscript.

Harman, G. (1973) *Thought*. Princeton, NJ: Princeton University Press.

Harman, G. (1974) Meaning and Semantics. In M. R. Munitz and P. K. Unger (eds), *Semantics and Philosophy*. New York: New York University Press, pp. 1–16.

Harman, G. (1982) Conceptual Role Semantics. *Notre Dame Journal of Formal Logic*, 20: 242–256.

Harman, G. (1986) *Change in View*. Cambridge, MA: Bradford Books/MIT Press.

Jeffrey, R. (1983) The First Definneti Lecture. Unpublished manuscript.

Kaplan, D. (1989) Demonstratives, in J. Almog, J. Perry, and H. Wettstein (eds.), *Themes from Kaplan*, (New York: Oxford University Press), 481–504.

Kripke, S. (1972) *Naming and Necessity*. In D. Davidson & G. Harman (eds), *Semantics of Natural Languages*. Dordrecht: Reidel, pp. 253–355.

Kripke, S. (1979) A Puzzle about Belief. In Margalit (ed.), *Meaning and Use*. Dordrecht: Reidel, pp. 239–283.

Lepore, E. (1982a) Truth and Inference. *Erkenntnis* 18: 379–395.

Lepore, E. (1982b) In Defense of Davidson. *Linguistics and Philosophy* 5: 277–294.

Lepore, E. (1983) What Modal Theoretical Semantics Cannot Do. *Synthese* 54: 167–187. [Chapter 3 this volume].

Lepore, E. & Loewer, B. (1981) Translational Semantics. *Synthese* 48: 121–133. [Chapter 1 this volume]

Lepore, E. & Loewer, B. (1983) Three Trivial Truth Theories. *Canadian Journal of Philosophy* 13/3: 433–47. [Chapter 2 this volume]

Lepore, E. & Loewer, B. (1986) Solipsistic Semantics, *Midwest Studies in Philosophy*. 10, 595–614, Chapter 9 this volume.

Loar, B. (1981) *Mind and Meaning*. Cambridge: Cambridge University Press.

Loewer, B. (1982) The Role of 'Conceptual Role Semantics'. *Notre Dame Journal of Formal Logic* 23: 26–39. [Chapter 4 this volume]

Lycan, W. (1981) Form. Function and Feel. *Journal of Philosophy* 78: 24–50.

Lycan, W. (1982) Toward a Homuncular Theory of Believing. *Cognition and Brain Theory* 4: 19–31.

Lycan, W. (1984) *Logical Form*. Boston, MA: MIT Press.

Lycan, W. (1985) Paradox of Naming. In B. K. Matilal and J. L. Shaw (eds), *Analytical Philosophy Comparative Perspective*, Dordrecht: Reidel, pp. 81–102.

McDowell, J. (1980) On the Sense and Reference of a Proper Name. In M. Platts (ed.), *Reference, Truth and Reality*. London: Routledge & Kegan Paul, pp. 111–130.

McGinn, C. (1982) The Structure of Content. In A. Woodfield (ed.) *Thought and Content*, Oxford: Oxford University Press: 207–258.

Perry, J. (1978) Frege on Demonstratives. *Philosophical Review* 59: 474–497.

Perry, J. (1979) The Problem of the Essential Indexical. *Nous* 12: 3–21.

Putnam, H. (1975) The Meaning of 'Meaning'. In H. Putnam, *Mind, Language, and Reality: Philosophical Papers*. Cambridge: Cambridge University Press, Vol. 2, pp. 131–171.

Putnam, H. (1986) Meaning Holism. In L. E. Hahn and P. A. Schalpp, *The Philosophy of W. V. Quine*, Open Court, pp. 405–431.

Putnam, H. (n.d) Computations and Interpretations [= CI]. Unpublished manuscript.

Sellars, W. (1956) Empiricism and the Philosophy of Mind. *Minnesota Studies in the Philosophy of Science*. Minnesota: Minnesota Press, Vol. 1, pp. 253–329.

Sellars, W. (1963) Some Reflections on Language Games. *Philosophy of Science* 21 (1951): 204–228.

Sellars, W. (1969) Language as Thought and as Communication. *Philosophy and Phenomenological Research* 29: 93–117.

Tarski, A. (1956) The Concept of Truth in Formalized Languages. In A. Tarski, *Logic, Semantics, and Metamathematics*, Oxford: Clarendon Press, pp. 152–278.

Taylor, B. (1980) Truth Theories for Indexical Languages. In M. Platts (ed.), *Reference, Truth and Reality*, London: Routledge & Kegan Paul, pp. 182–198.

White, S. (1982) Partial Character and the Language of Thought. *Pacific Philosophical Quarterly* 63: 347–65.

Zemach, E. (1976) Putnam's Theory of the Reference of Substance Terms, *Journal of Philosophy* 73/5: 116–127.

6

What Davidson Should Have Said (1989)

Ernest Lepore and Barry Loewer

It is natural to think, as many have, that the expressions, both simple and complex, of a language L possess meanings and that a theory of meaning for L should specify these meanings. Such a theory would, it is said, illuminate the semantics of L by showing how the meanings of its expressions are interrelated (for example, how the meanings of complex expressions depend on the meanings of their parts), and it would also illuminate *meaning*, at least for L, by specifying the extension of 'means' in L. Further, if, as many have claimed, understanding L consists at least in part in knowing, if only implicitly, the extension of the meaning relation for L, such a theory of meaning would be the core of a theory of understanding L. A first proposal concerning a meaning theory for L is that it issue in true theorems of the form:

(M) S means in L that *p*

for all sentences S of L. If we suppose that meanings are the referents of expressions of the form 'that *p*', then a theory (TM) which entailed all and only true instances of (M) for sentences of L would fix the extension of 'means in L' (and, presumably, given sufficient metalanguage apparatus, the theory would entail an explicit definition of 'means in L'). There is, however, a number of difficulties with this way of formulating a theory of meaning.

While instances of (M) specify the meanings of sentences of L, they obviously are inappropriate for specifying the meanings of subsentential expressions: '"and" means and' is ill-formed. Further, without having some way of specifying the meanings of subsentential expressions, it is difficult to see how we would be able to derive the true instances of (M), since the meanings of sentences depend on the meanings of their parts. Another problem is created by the apparent intensionality of 'means in L that'. A theory which entails instances of (M) presumably will be formulated in an intensional language. But the logics of intensional (as well as of extensional) languages for which semantics have been developed contain a principle which allows for the substitution of logically equivalent sentences. So, if (TM) entails 'S means that *p*', it will also entail 'S means that $p^{\star}$', where '*p*' and '$p^{\star}$' are logically equivalent. But logically equivalent sentences need not possess the same meaning.

Finally, perhaps for the reasons just mentioned, no one has produced a theory which entails all the instances of (M) for even a simple first-order language.

Another approach to specifying meanings is 'tranlational semantics'. Instead of instances of (M), a translation theory produces instances of:

(TR) e in L means the same as $e^\star$ in $L^\star$

A theory entailing instances of (TR) does not face the same problems as (TM), since 'means the same as' is purely extensional and its extension is not restricted to sentences. But such a theory may not tell us much about the meanings of the expressions of L. In particular, someone might know all the instances of (TR) for L and yet not understand L. Thus someone might be able to match Italian expressions with their Russian translations, say, on the basis of lexographic and syntactic rules, and yet not understand the expressions of either language. (For further discussion and criticism of this approach, see Chapter 1 of this volume.)

Davidson has suggested another way of formulating a theory of meaning for L, which seems to avoid the pitfalls both of the translational and of the meaning approach. Instead of trying to produce a theory which specifies the extension of 'means that' or 'means the same as', Davidson asks what we want a theory of meaning to do. His answer is that a theory of meaning should specify information such that if someone had this information he would be in a position to understand L, or at least well on his way to understanding L (Davidson, 1967; see also, 1973, and 1974). He then claims that a theory of truth for L fits this description. A theory of truth for L issues in theorems of the form:

(T) S is true in L iff p

for each sentence of L. Truth theories are purely extensional; they apply via the satisfaction clauses to subsentential expressions; and we know, at least for formal languages, how to construct them. The information they specify, unlike that of translation theories, does seem relevant to understanding L. An instance of (T) says under what conditions S is true, and not merely that S has the same truth value as a sentence in some (other) language. As Tarski observed, a true theory for L fixes the extension of 'is true in L'. But how does it illuminate meaning in L? There are two claims, each related to Davidson's characterization of the goal of an adequate theory of meaning:

[I] The truth theory shows how the truth conditions of complex sentences depend on the truth conditions and other semantic features of constituent expressions. Thus, a truth theory (TT) entails, for example, ' "S & S*" is true iff p and q' by deriving it from 'S is true iff p' and 'S* is true iff q' and an axiom '$(\forall x)(\forall y)$ $(x$-'&'-$y)$ is true iff x is true and y is true)'.

[II] A truth theory (TT) issues in *true* theories of the form (T). Although such 'T-sentences' do not *say* that S means that p, Davidson suggests that in fact an adequate truth theory will entail that 'S is true iff p' iff S means that p.

If condition [II] is satisfied by (TT), then we will have succeeded in pairing each sentence of L not only with its truth condition, but with what we might call its meaning condition. We need to address the questions whether there are such truth theories for L, and, if so, which theories count as such. This divides into the double question whether, if (TT) is a true theory for L, (TT) will satisfy:

(1) If S means that p, then (TT) will entail 'S is true iff p'.
(2) If (TT) entails 'S is true iff p', then S means that p.

Tarski stipulated that a theory counts as a truth theory for L only if it entails an instance of (T) for each S of L in which 'p' is a translation of S. But, as Davidson observes, we do not want to make this stipulation in formulating a theory of meaning, since we would like the fact that 'p' translates S to emerge from the conditions placed on the truth theory. Minus the translation condition, the constraints on an adequate truth theory for L are:

(i)   that it entail a T-sentence for each S in L;
(ii) that it be true; and
(iii) that it be finitely axiomatizable.[1]

Clearly, not every truth theory for L satisfying (i)–(iii) will satisfy (1). For example, it is plausible that, if there is any truth theory for a fragment of English containing 'Snow is white', 'Grass is green', and kindred sentences, then there will be a truth theory which entails ' "Snow is white' is true iff grass is green' (but not ' "Snow is white" is true iff snow is white'). In 'Truth and Meaning', Davidson suggested that the demands on a truth theory—that it be complete (or that it entail a T-sentence for each sentence of L) and that it be true (for example, that it may not contain ' "snow" refers to grass')— make it impossible for there to be such a theory. He was wrong.

Suppose that (TT) is a truth theory which satisfies (1). Presumably there are such theories. From it, though, we can obtain another truth theory (TT'), which satifies (i)–(iii), but fails to satisfy (1). Just take some predicate clause, for example, '$(\forall x)(x$ satisfies 'is wet' iff x is wet)' and replace it with '$(\forall x)(x$ satisfies 'is wet' iff x is wet*)', where 'is wet*' has as its extension wet things in the actual world and dry things in other worlds.[2]

The question naturally arises: Are there further constraints that can be placed on a truth theory to guarantee the satisfaction of (1)? Before canvassing some attempts to

---

[1] We do not want to discuss or motivate condition (iii) in this chapter. See Chapter 2 in this volume and Lepore, 1982.

[2] It is not clear that this will work. Suppose the language has counterfactuals. How will an absolute truth theory deal with these? It might be that if we have counterfactuals, then our cooked-up clause will yield false T-sentences, violating condition (ii). In any case, there obviously are theories which satisfy (i)–(iii), but fail to satisfy (1). For example, a theory which entails 'S is true iff 2+2=4' or 'S is true iff 2+2=5' (contingent upon whether S is true or false). Among such theories there is one which has just true consequences. Of course, one would not know that one had such a theory unless one knew the truth values for all sentences of L.

answer this question, we should see that only certain kinds of constraints are admissible. Presumably, if there are any truth theories for L, there is one which satisfies (1). But no condition which already presupposes understanding L is appropriate if we want the theory to illuminate meaning. Suppose we found further appropriate conditions C which guarantee satisfaction of (1). Such a theory would pair each S with a sentence '*p*', which gives its meaning, although it will be weaker than a meaning theory since it will not *say* that S means that *p*. Still, we can expect that such a theory, if we should find one, will illuminate 'means that in L'.

There are two suggestions we know for constraining truth theories so that they satisfy (1). The first one says that the T-theorems should be law-like (Davidson, 1976) This removes certain theories which violate (1). Thus, ' "Snow is white" is true iff snow is white' is a law, while ' "Snow is white" is true iff grass is green' is not a law. Presumably the underlying thought is that it is a matter of accident that the second is true. But this will not suffice to guarantee that we derive all the T-sentences we want. For example, a good theory will entail ' "Water is wet" is true iff water is wet'. But if this is a law, then ' "Water is wet" is true iff H_2O is wet' also is (seems to be) a law. Therefore, there will be an adequate truth theory which implies the latter, not the former, and accordingly fails to satisfy (1).[3]

A second suggestion is that the truth theory should make certain predictions concerning the speakers of L (Davidson, 1967, 1974; Evans and McDowell, 1976; Platts, 1979). These predictions concern what sentences the speakers of L hold true under various circumstances. In that way, the theory can be used to predict correct beliefs on the basis of sentences held to be true. We will imagine that we have identified the sentences of another speaker's language and those which that speaker holds to be true. We test a theory of truth for his language in the following way: if the speaker holds S true and the truth theory that we are testing entails that S is true if and only if *p*, then we will assume he believes that *p*. Different theories will predict he has different beliefs. Thus, a (TT) which entailed the wrong sentences would predict (together with reasonable accounts of belief formation) that after spending a winter in New York City an English speaker would no longer hold it to be true that 'Snow is white', but he would still hold 'grass is green' to be true. (The latter will be reversed after his spending a summer in Los Angeles during a drought.)

It is not clear to us whether either condition or the two together (they are obviously closely related) will filter out all truth theories except those that satisfy (1). One troubling thought for this second line is that, if we do not have access to a person's beliefs independently of access to his language, as is suggested by some of Davidson's writings, then there may be alternative truth theories and belief assignments (or accounts of belief formation) which together make exactly the same predictions concerning which sentences the person holds true or would hold true. If so, then

---

[3] Someone might argue that, although both statements are laws, only the first is a *semantic* law. This suggestion is vacuous unless a non-question-begging characterization of semantic law is provided.

this condition will not do as much winnowing as anticipated. However, a defender might respond that it was wrong to expect this.

Although we are not sure whether any non-question-begging constraint will yield truth theories which satisfy (1), we are certain that no constraint of any sort will lead to a satisfaction of (2). The difficulty is obvious and one we already have mentioned in our discussion of the troubles with any effort to construct a meaning theory. It is the problem of substitutivity of logical equivalences. If (TT) entails that S is true iff p, and if p and $p^\star$ are logically equivalent, then (TT) will entail S is true iff $p^\star$. Indeed, the problem is even worse; if (TT) entails anything R, and it entails that S is true iff p, then it will entail that S is true iff p & R. So we will derive theorems like ' "There is a fire in the building" is true iff there is a fire in the building and (q or not q)'. Put simply, as long as we are using normal logical apparatus (the theory is formulated in a first-order language or some extension thereof), then satisfying (2) is out of the question. And any tinkering with the logic of the truth theory to prevent it from implying 'bad' T-sentences would also prevent it from implying 'good' T-sentences.

How bad is this? Should we then abandon the idea that a theory of truth is the core of a theory of meaning? No, but we should abandon the idea that a theory of truth *serves* as a theory of meaning by producing T-sentences which simulate M-sentences. Once we return to Davidson's suggestion that a theory of meaning for L should specify information sufficient for understanding L, we should see why this is so. Curiously, neither Davidson nor others who have advanced or seconded his suggestion (including Davidson's Oxford followers) have ever quite articulated exactly how knowledge of a truth theory (even one satisfying (1) and (2), if there were one) would yield a theory of understanding L. Perhaps it was thought that, if a (TT) satisfied (1) and (2), then it would be obvious how, since knowing (TT) would amount to knowing the meaning conditions of sentences of L. But what has this got to do with understanding language? To understand a language involves at least this much: to be warranted in drawing (or acting as if one were drawing) inferences characteristic of understanding.

Suppose Donald and Luca are in a hotel in New York waiting for the mail. Mary enters the room and utters 'There is a fire in the building'. Donald understands English and Luca does not. On this basis, all other things being equal, we would expect that Donald leaves the building, while Luca just sits there. If in fact that happens, it would be natural to explain Donald's behaviour by noting that upon hearing Mary he acquired the belief that there is a fire in the building. Donald's belief (and his ensuing behaviour) is warranted, in part, by the fact that he understands English. The exact psychology of Donald's belief acquisition is a matter for psychology and we do not intend to address that matter here. Our conception of truth-conditional semantics emphasizes its *epistemological*, rather than its psychological, role (see Chapter 5 in the present volume). If we want to specify knowledge which, if Donald had it, it, together with other things Donald (and for that matter Luca) knows, would *justify* his belief, i.e.

the knowledge that 'there is a fire in the building' is true if and only if there is a fire in the building is the natural candidate.[4] The following reasoning makes this explicit:

Mary's utterance 'There is a fire in the building' is true.
Mary's utterance is true only if there is a fire in the building.
So, there is a fire in the building.

The aim of producing a theory which satisfies (1) and (2) is not an appropriate aim for a theory of meaning. This aim should not be to *define* or *analyze* meaning. Rather, it should be to produce a theory which explains the phenomena for which we invoke the concept of meaning. Here the phenomenon is language understanding. Davidson never claimed that the bi-conditionals issued in meanings. He claimed, rather, that knowing *certain* bi-conditionals, those implied by an adequate truth theory, suffices for understanding. Nothing is upset by knowing additional true bi-conditionals. For example, someone who knows the above would be in a position to conclude that there is a fire in the building, and (q or not q), for any q. But there is still a problem.

Even if one knew a truth theory for Mary's language which satisfies (1), and knew that Mary believes all the T-consequences of this theory (or believes the theory), this would be insufficient for understanding it. When we know that Mary holds true 'There is a fire in the building', say, because she uttered it, we are then in a position to justify our beliefs that there is a fire in the building, assumming she is reliable and sincere. But we cannot conclude that Mary *said* that there is a fire in the building. The real trouble with a truth theory as a theory of understanding is that its T-sentences are too weak to warrant inferences to what someone said (asked, commanded) on the basis of what he uttered. Thus, truth theories qua theories of understanding become unhinged just where we demand them to be most secure, namely in accounting for inferences to what others say on the basis of what they utter.

We return to New York. Suppose that, when Mary uttered 'There is a fire in the building', only Donald was within earshot. Provided Donald understands Mary's language, then Donald would, in 'normal' circumstances, be warranted in believing that Mary said that there is a fire in the building. Indeed, aren't these sorts of belief constituitive of understanding? Should Donald want to inform Harry of Mary's utterance, i.e. her opinion, he can take recourse either to direct or indirect quotation. Either way, Donald treats himself as a conduit for information from Mary to Harry. Should Donald utter 'Mary said 'There is a fire in the building', then, if Harry understands English, minimally he would be warranted in believing that Mary uttered a sentence of her language of the same phonological type as the sentence uttered by Donald. Harry might further conclude that Mary said that there is a fire in the building. How is this possible? Many writers have argued there can be no answer within the Davidsonian 'extensional' framework to this question. Indirect quotation is

---

[4] Both Donald and Luca could know that Mary asserted 'there is a fire in the building', and even that her utterance is true. Evidence for an utterance being an assertion and for it being true need not be semantic.

an 'intensional' notion, so the folklore goes, and truth theories are extensional animals. How could knowledge of just the conditions under which sentences are, as a matter of fact, true ever support inferences inside apparently intensional contexts like 'Mary said that [. . .]?' We quickly restate the objection and answer it. Our answer has important consequences for theories of interpretation.

Foster (1976), Loar (1976), Wallace (1976), Evans and McDowell (1976), among others, objected to Davidson that the trouble with a truth theory as a theory of meaning is that its T-sentences are too weak. It is this, they argue, that allows theories (or theorems) in which S does not mean that *p*. Furthermore, it might be thought, it is just this weakness that prevents one from concluding what a speaker *says* when he produces an utterance. It looks like 'A uttered S and S means that *p*' does entail 'S said that p'. So there seem to be two distinct issues, easy to conflate:

(i) adding further conditions on a *truth theory* so as to produce one that satisfies the MC, toss out the 'bad' ones; and

(ii) determining how to use the truth theory to obtain conclusions like 'A said that p'.

(i) Issue cannot, as a matter of logic, be solved for the reason we advanced earlier.[5]

However, as we also argued, the objection that a bi-conditional like ' "There is a fire in the building" is true iff there is a fire in the building and (*q* or not *q*)' does not provide the meaning of 'There is a fire in the building' is misdirected. Issue (ii) is a different objection and it is not misdirected: (utterances of) 'There is a fire in the building' and 'there is a fire in the building and (*q* or not *q*)' do not *say* the same thing. Any competent speaker of English knows this. So the defenders of truth-theoretic approaches to meaning are left with the problem of discovering a way of representing information which can justify conclusions concerning what is said that are compatible with the truth-theoretic approach. As we will now show, Davidson's paratactic account, somewhat unexpectedly, provides a solution to this problem.

Whenever Mary utters 'there is a fire in the building', she produces an utterance which is the same in content (purport, import) as many other utterances of the English sentence 'there is a fire in the building'. But then it would follow that there is an utterance *u* such that Mary produced *u* and *u* is the same in content as 'there is a fire in the building'. This is, according to paratactic account (Davidson, 1968, 1969), the correct semantic account for an utterance 'Mary said there is a fire in the building', which happens to be our target sentence. On the paratactic account, the fact that Mary uttered 'there is a fire in the

---

[5] This has little to do with truth theories *per se*; rather it has to do with the truth theory being a *theory*. Suppose we add a modal operator, both as Foster (1976) and Wallace (1976) suggest, to strengthen the bi-conditionals which issue from an adequate truth theory. For example, Wallace suggests that we add as a primitive the operator 'it is a matter of meaning alone'. Therefore, according to him, our target sentences are of the form: it is a matter of meaning alone that S is true iff *p*. But how, if his theory has standard logical principles (and how can it not?), is Wallace going to prevent the derivation of false theorems of the form: it is a matter of meaning alone that S is true iff *p* and (*q* or not *q*)?

building' and the fact that her utterance bears the *samesay* relation to (another) English speaker's utterance of 'there is a fire in the building' imply that Mary said that there is a fire in the building.[6] If we assume, then, that 'that' refers to an utterance of 'there is a fire in the building', then, if the premises are true, the conclusion is true. The conclusion just is 'Mary said that'. The exact nature of this inference requires us to discuss the logic of demonstratives—something that space limitations prevent us from doing here (but see the next chapter in this volume). However, even without this discussion, we have said enough to show how we can overcome perhaps the major objection to Davidson's theory, namely that the theory is too weak to support inferences from 'Mary uttered "There is a fire in the building"' to 'Mary said that there is a fire in the building'. Sometimes Davidson says that knowledge of a truth theory for a language L is *sufficient* for understanding L (1967, 1973, 1974) Later (in his reply to Foster, 1976) he modified this by adding that one also must know that the truth theory meets certain empirical and formal constraints. The critics were right inasmuch as neither knowledge of a truth theory nor knowledge that the truth theory is an empirically adequate truth theory can bridge the gap to what is said (even though it effectively bridges the gap to knowledge that there is a fire in the building). What we have just shown is that everyone was looking in the wrong place. It is not (and cannot be) the truth theory, no matter how it is constrained or what is known about it, that bridges the gap between heard utterance and ascribed assertion. What Davidson should have said is that one needs also to know the samesay relation, or some proper subset of this relation, for the language. How much does this add to Davidson's overall project of stating in extensional terms what it is, in what one knows, that enables one to understand a language? Not much. One needs to know the samesay relationship anyway for interpreting indirect discourse, if Davidson's own extensional account of this sort of discourse is correct.[7]

It is important to realize that we are not saying that all one needs to know in order to understand a language is the samesay relation for this language. After all, one *can know* that one sentence S samesays another without understanding S (indeed, without understanding either). Thus, on the paratactic account, one could be in position to

---

[6] We need to be careful here. Strictly speaking, the two premises do not logically imply the conclusion, since there are contexts in which the premises are true but the conclusion is false—for example in which 'that' refers to some other utterance.

[7] The question naturally arises how one can know when two utterances samesay each other. That is a hard question to answer. It is recognized by speakers of any language that no two utterances are likely ever to carry exactly the same information to all audiences. Because of this, speakers treat the samesay relation as *flexible* and *pragmatic*. Whether a claim that two utterances samesay each other is counted as true among speakers of a language depends on pragmatic matters, including the point of making the claim and its intended audience. Sometimes a great deal of leeway is allowed. Indeed, since speakers recognize that others may use somewhat different dialects and that over time the import of words change, they sometimes get into discussions and debates concerning whether two utterances samesay each other. But, for most matters, there is sufficient agreement and speakers master the art of recognizing when two utterances do or do not stand in the samesay relation to each other. So, even if philosophers studying the speakers of a language cannot provide necessary and sufficient conditions for the relation in, say, physicalistic, or even psychological or semantic terms, they and we can be confident that the practices of the speakers suffice to fix (with some margin of vagueness and indeterminacy) the predicate's extension.

conclude 'Mary said that *p*' without *understanding* what Mary said. But here is exactly where knowledge of the truth theory enters. Our claim is not that speakers need *only* know which utterances samesay each other, they must also know the truth conditions for these utterances. They can then employ this knowledge together with knowledge of the extension of the samesay relation to conclude, in accordance with patterns of inference now familiar, that Mary said something which is true if and only if there is a fire in the building.

In conclusion, we say that, to understand a language, it suffices to know:

(a) all the bi-conditionals entailed by an adequate theory; and,
(b) for any two utterances of sentences of the language, whether or not they samesay each other.

Notice that this does not require knowing theories which entail the specific knowledge in (a) and (b). Both kinds of knowledge seem to be infinite, so it may be impossible to *specify* either without devising a theory. That is not our concern here. Our concern was to show in what sense a truth theory is *not* too weak to serve as a theory of meaning and in what sense it is.

We argued that both the critical literature and Davidson's response to this literature have been misguided. Critics have been misguided because they have not been clear about what it is that a theory of meaning is supposed to do. Some critics have been misguided because they have thought that a theory of meaning for a language L is supposed to pair each sentence of L with its meaning. No truth theory can do this. Both Davidson and his critics seem to have been misguided in thinking that, by adding further conditions on an empirically adequate truth theory, we can derive what was asserted. We showed that Davidson had, available to him though he apparently failed to see it as such, a reply to his critics in his account of the semantics for indirect discourse reports. On this account, 'Mary said that there is a fire in the building' and 'Mary made an utterance which samesays my utterance of There is a fire in the building' are equivalent. Everything stands on this, and obviously someone might object—and indeed some have objected here. Whether this account is itself acceptable is the topic of the next chapter.

Acknowledgments

An earlier draft of this chapter was presented at a Symposium on Donald Davidson's philosophy in Gornja Radgona, Yugoslavia, and Bad Radkersburg, Austria, November 12, 1988, and at the University of Minnesota. We would like to thank both groups for the opportunity they offered us to present our work. We would also like to thank Michael Root, John Wallace, and especially Donald Davidson for their comments.

References

Davidson, D. (1967) Truth and Meaning. *Synthese* 17: 304–323; reprinted in Davidson (2001).

Davidson, D. (1968) On Saying That. *Synthese* 19: 130–146; reprinted in Davidson (2001).

Davidson, D. (1969) True to the Facts. *Journal of Philosophy* 66: 748–764; reprinted in Davidson (2001).

Davidson, D. (1973b) Radical Interpretation. *Dialectica* 27: 314–328, reprinted in Davidson (2001) [= RI].

Davidson, D. (1974) Belief and the Basis of Meaning. *Synthese* 27: 309–323; reprinted in Davidson (2001) [= BBM].

Davidson, D. (1976) Reply to Foster. In G. Evans and J. McDowell (eds), *Truth and Meaning: Essays in Semantics*, Oxford: Clarendon Press, pp. 33–41; reprinted in Davidson (2001).

Davidson, D. (2001) *Inquiries into Truth and Interpretation*, 2nd edn, Oxford: Oxford University Press.

Evans, G., and J. McDowell (eds) (1976) *Truth and Meaning*. Oxford: Oxford University Press.

Foster, J. A. (1976) Meaning and Truth Theory. In Evans and McDowell, pp. 1–32.

Lepore, E. (1982) In Defense of Davidson. *Linguistics and Philosophy* 5: 279–94.

Lepore, E. and B. Loewer (1981) Translational Semantics. *Synthese* 48: 121–33, Chapter 1 this volume.

Lepore, E. and B. Loewer (1983) Three Trivial Truth Theories. *Canadian Journal of Philosophy* 13/3: 433–447. [Chapter 2 this volume]

Lepore, E. and B. Loewer (1986) Dual Aspect Semantics. In Ernest Lepore (ed.), *New Directions in Semantics*, Academic Press, pp. 83–112. [Chapter 5 this volume]

Lepore, E. and B. Loewer (1989) You Can Say That Again. *Midwest Studies in Philosophy* 12: 338–356. [Chapter 7 this volume]

Loar, B. (1976) Two Theories of Meaning. In Evans and McDowell, pp. 138–161.

Platts, M. (1979) *Ways of Meaning*. London: Routledge and Kegan Paul.

Wallace, J. (1976) 'On Davidson's Analysis of Indirect Discourse'. Unpublished manuscript.

7

You Can Say *That* Again (1989)

Ernest Lepore and Barry Loewer

It has been two decades since the publication of Davidson's twin papers, 'Truth and Meaning' (1967) and 'On Saying That' (1968). The first proposed that a Tarskian truth theory for a language L can be (the heart of) a theory of meaning for L. The second proposed a radically novel approach—the paratactic account—to the logical form of indirect discourse. The two proposals are related in a number of ways. The paratactic account claims to show how to construct a truth theory for languages containing indirect discourse and propositional attitude reports. The truth-theoretic account of meaning provides motivation and support for the paratactic account. And, somewhat surprisingly, the paratactic account provides a way to get around an objection that many have seen as fatal to Davidson's truth-theoretic account of meaning. Indeed, the proposals are so interconnected that, if one fails, it is likely that the other cannot be successfully defended. Our primary aim in this chapter is to motivate the paratactic account and to defend it from certain widely known criticisms. Along the way, we will spell out various connections between the paratactic account and the truth-theoretic account of meaning, and we will explain how the former can be used to support the latter.

1.

One aim of a theory of meaning for a language L is to specify information concerning L which, if someone possessed it, would enable him to interpret utterances of speakers of L. But what is it to interpret another's utterances? To answer this question, consider two people: one understands Italian, the other does not. Suppose that both of them hear Andrea utter 'Il gatto siede dietro al forno'. Each one may know that Andrea's utterance is an assertion, and that it is true.[1] But only the one who understands Italian, other things being equal, will be justified in believing, on the basis of hearing Andrea,

---

[1] Evidence for an utterance being an assertion and for a particular utterance being true need not be semantic. For example, one might know that certain sentential forms and tones of voice indicate assertion and that a particular person's assertions in certain circumstances are generally true without knowing what his assertions mean.

that the cat sits behind the oven. What semantic information would justify this belief? The inevitable answer is this: *Andrea's utterance is true only if the cat sits behind the oven.* The following reasoning makes this explicit:

(1) Andrea's utterance 'Il gatto siede dietro al forno' is true.

(2) Andrea's utterance is true *only if* the cat sits behind the oven.

(3) So, the cat sits behind the oven.

Furthermore, the information that

(2¢) Andrea's utterance is true *if* the cat sits behind the oven

can justify Andrea's uttering 'Il gatto siede dietro al forno', supposing that Andrea wants to produce an utterance which is true if a cat sits behind an oven.[2]

An Italian speaker is also in a position to learn from utterances something about the beliefs of Italian speakers—namely, in our example, that Andrea believes that the cat sits behind the oven. Knowledge of truth conditions can play a role in this justification as well:

(4) Andrea believes that his utterance of 'Il gatto siede dietro al forno' is true.

(5) Andrea believes that his utterance is true iff the cat sits behind the oven.

(6) So, Andrea believes that the cat sits behind the oven.

(4) is based on the presumption of sincerity. (6) does not logically follow from (4) and (5) (unlike the first inference, where (1) and (2) imply (3)). But in this case it is overwhelmingly plausible that (6) is true, given (4) and (5). Why else would Andrea have made the utterance? These observations provide straightforward support for the truism that understanding a language involves knowledge of truth conditions.

A truth theory for a language L assigns to the infinitely many possible utterances of L truth conditions based (in part) on the syntax of the sentences uttered, and thus systematizes the truth conditions of the sentences of L. The information embodied in such a theory, as we just saw, provides justifications for (some of) the beliefs that a competent language speaker can acquire upon hearing the assertive utterances of speakers of L. It also provides (partial) justifications for some of her own utterances. A truth theory for L will also specify a great deal of other information concerning L, information concerning relations among the truth conditions of sentences of L (for example, logical relations) and information concerning reference relations. But our interest here is in a truth theory's role as a systemization of the truth conditions for utterances of L.

---

[2] Andrea may have various reasons for wanting to produce an utterance which is true if the cat sits behind the oven. One such reason is that he wants his audience to acquire the belief that the cat sits behind the oven, and he believes that his producing an utterance which is true if the cat sits behind the oven is likely to accomplish this. The reasonableness of the latter belief may involve his believing that the audience believes 'Il gatto siede dietro al forno' to be true if and only if the cat sits behind the oven.

YOU CAN SAY *THAT* AGAIN 99

It seems to us uncontroversial that speakers of a language know the truth conditions of sentences of their language. But whether speakers actually know a truth theory and whether such knowledge (or even knowledge of truth conditions) is causally involved in the production and understanding of utterances are psychological questions we do not intend to address here. Our conception of truth-conditional semantics emphasizes its *epistemological* rather than its psychological role. It is clear that a speaker of Italian will be in a position to acquire and be justified in acquiring a great many beliefs about the world and about the mental states of his fellow speakers in virtue of his understanding Italian. A semantic theory of Italian, in our view, should systematize information which a speaker possesses and which justifies the beliefs acquired in the ways we have outlined.

A competent speaker of Italian will also be in a position to acquire, from his belief that Andrea assertively uttered ' "Il gatto siede dietro al forno", the belief that Andrea said that the cat sits behind the oven. The justifying inference might look like this:

(7) Andrea assertively utters ' "Il gatto siede dietro al forno".
(8) [Some premise stating semantic information.]
(9) So, Andrea said that the cat sits behind the oven.

What should premise (8) be? It is clear that ' "Il gatto siede dietro al forno" is true iff the cat sits behind the oven' is too weak. Supposing the cat sits behind the oven, then 'Il gatto siede dietro al forno' is true iff snow is white. This certainly does not justify the conclusion that Andrea said that snow is white. And not even knowing that *Andrea* knows his utterance to be true iff the cat sits behind the oven—or that this is common knowledge among all the speakers of Italian—will suffice. The 'obvious' missing premise that would suffice is that Andrea's utterance *means that* the cat sits behind the oven. But, of course, one main point of 'Truth and Meaning' is to replace the view that a theory of meaning for L entails sentences of the form 'S means that *p*' (M-sentences) with the view that it entails sentences of the form 'S is true iff *p*' (T-sentences). Davidson made this claim because:

(i) he thought that, due to the intensionality of 'means that', the logical machinery involved in proving M-sentences to be correct would certainly be vastly more complex, and might ultimately be unobtainable; and

(ii) he thought that a theory which issues in T-sentences will accomplish what we can rightfully expect from a theory of meaning.

We do not want to discuss here Davidson's doubts about theories issuing in M-sentences. But Davidson was mistaken in the second claim, since, as we have just noted, a truth theory falls short in that it fails to justify beliefs concerning what is said. This does not seem to be a minor failing, since knowing that A said that *p* on the basis of his uttering S is central to understanding S. So defenders of truth-theoretic approaches to meaning are left with the problem of discovering a way of representing information which can justify conclusions concerning what is said which are compatible with the truth-theoretic approach. As we will show, the paratactic account, somewhat unexpectedly, provides an end run around this problem.[3]

Our strategy for explaining the paratactic account is to describe a population P which speaks a language L for which we stipulate that the account is correct. Let us suppose that L is a first-order language containing names (of people, places, things, times, and (utterances of) sentences of L) and predicates. Among these predicates are three two-place predicates: $U(a,u)$, $S(u,u^\star)$, $SS(u,u^\star)$. The first relates a person to an utterance, the second and third relate one utterance to another. (We will explain them later.) L also contains indexicals, tenses, and demonstratives. We will suppose that the sentences of L have truth conditions and that the speakers of L all know the same truth theory. This will enable, for example, one member of P to learn from another that snow is white when he hears him assertively utter 'Snow is white'—assuming, of course, that he believes the utterance is true. The presence of indexicals and demonstratives in L enables speakers of L to use a single sentence to convey different information. The information conveyed depends on the utterance's context in systematic ways. For example, an utterance of 'I am hungry' is true if and only if the speaker who delivers it is hungry at the time of the utterance. The presence of indexicals, demonstratives, and tenses causes certain complications in the form of T-sentences. Suppose that one member of P, Arabella, hears another, Barbarella, assertively utter 'I am hungry'. Then Arabella could reason as follows:

(10) Barbarella's utterance of 'I am hungry' is true.
(11) (u) (x) (if u is an utterance of 'I am hungry' by x, then u is true iff x is hungry)
(12) So, Barbarella is hungry.

The truth condition associated with utterances of the syntactic type 'I am hungry' is a generalization. Someone who knows it is in a position to employ information she possesses concerning the utterance so as to draw conclusions about who is hungry. If Arabella did not know that the utterance was made by Barbarella, but, say, that it was made by that woman over there, then she would not be justified in concluding that Barbarella is hungry, but she would be justified in concluding that the woman over there is hungry. Minimally, she will always be justified in the belief that whoever made the utterance is hungry. The truth conditions for a sentence containing a demonstrative are also generalizations:

(13) (u) (x) (if u is an utterance of 'That is F', in which 'that' demonstrates x, then u is true iff x is F)

---

[3] There are two problems concerning truth-theoretic proposals that sometimes are conflated. There are truth theories for L which entail T-sentences where S and 'p' differ in meaning. In fact, any (interesting) truth theory will have this feature. The second problem is that 'S is true iff p' does not *say* that S means that p. Davidson tried to remedy the first problem by requiring that truth theories meet certain empirical constraints on the theory of interpretation and the second by adding that an interpreter must know not just that S is true iff p, but that 'S is true iff p' is entailed by a theory meeting those constraints. We have argued in Chapter 6 that Davidson's remedies are inadequate.

A truth theory for languages which contain indexicals and demonstratives will imply, for each sentence type S of L, a theorem of the form:

(T⋆) (*u*) (*c*) (if *u* is an utterance of S in context *c*, then *u* is true in L iff *p*(*f*(*c*), *g*(*c*), . . .)),

where '*f*', '*g*', and so on are functions which assign to a context certain of its features— for example the speaker of *c*, the time of *c*, and so on. A point worth noting is that the metalanguage in which a truth theory for an indexical language like L is constructed may itself contain indexicals and demonstratives. In that case we can represent in the metalanguage an inference from:

(14) Arabella's utterance of 'I am hungry' is true.
(15) She is hungry, demonstrating Barbarella.

We will not construct a theory for L, but we will assume that the speakers of L all know the same truth theory for L. This knowledge will enable a speaker of L to obtain information from the utterances of other speakers of L, information not necessarily obtainable by those who do not understand L.

Speakers of L have various ways of referring to sentences and utterances, including corner quotes, definite descriptions, names; but an especially convenient means is to refer demonstratively. Demonstrative reference to utterances is convenient, since one can produce an utterance and then demonstrate it. The predicates 'U(*a*,*u*)' and 'S(*u*,*u*⋆)' work as follows:

'U(*a*,*u*)' is true of a person *a* and utterance *u* iff *a* is the utterer of *u*.
'S(*u*,*u*⋆)' is true of utterances *u* and *u*⋆ iff *u* and *u*⋆ are utterances of the same sentence.

Suppose Arabella assertively utters:

(16) (E*u*) (U(Barbarella,*u*) & S(*u*,that)) [Snow is white.]

Her utterance of (16) consists of two utterances: an asserted utterance and an utterance demonstrated by this asserted utterance. (The demonstrated utterance is of the sentence in square brackets in (16).) The first utterance is true just in case Barbarella made an utterance of the same sentence type as the second utterance.

Utterances like (16) potentially carry a great deal of information which can be unlocked by someone who knows a truth theory for L. The first part of (16) has generalized truth conditions (17):

(17) (*v*) (*c*) (if *v* is an utterance of '(E*u*) (U(Barbarella,*u*) & S(*u*,that))', then *v* is true in *c* iff Barbarella made an utterance which is of the same sentence type as *d*(*c*),

where '$d(c)$' is the utterance demonstrated in context c. If Cinderella believes that Arabella's utterance is true and knows its truth conditions, then she will be licensed to conclude that Barbarella made an utterance of 'Snow is white'. The soundness of the inference requires Cinderella to recognize that the demonstrated utterance is of 'Snow is white'; but, since such an utterance has *just* been produced and since she is a speaker of L, we may assume that her phonological and syntactic knowledge warrants this belief. Although it is conventional in L that the demonstrated utterance is, other things being equal, not an assertion, Cinderella may be in a position to extract information from it. Suppose that she believes that Barbarella, as well as Arabella, is reliable. She could then reason as follows:

(18) (Eu) $(U(\text{Barbarella},u)$ & $S(u,\text{that}))$ [from previous inference]
(19) $U(\text{Barbarella},u^\star)$ & $S(u^\star,\text{that})$ [by existential generalization]
(20) $u^\star$ is true [Barbarella's reliability]
(21) That is, true [utterances of the same non-indexical sentences have the same truth value]
(22) 'Snow is white' is true [that is an utterance of 'Snow is white']
(23) 'Snow is white' is true iff snow is white [from the truth theory]
(24) So, snow is white.

Observe the role played by Arabella's having produced, and then demonstrated, an utterance in this inference. The demonstrated utterance 'stands in' for Barbarella's utterance, so she can apply her semantic knowledge to it and learn from it, for example, that snow is white.

It should be clear that (16) plays a role similar to a direct discourse report in English. L also contains a device similar to English *indirect* discourse. Such a device is needed for at least the following reason. We saw that Cinderella could conclude from Barbarella's reliability and from the truth conditions for the second part of (16) that snow is white. Her reasoning involved the assumption that, if $S(u,u^\star)$ and u is true, then $u^\star$ is also true. But generally this is false. Utterances of the same sentence, for example, 'I am hungry', obviously may differ in truth value. On the other hand, utterances of sentences of different syntactic types can agree in import; that is, in the information which someone who understands the utterances can obtain in virtue of his understanding. When Barbarella utters 'I am hungry', Arabella can produce an utterance with the same import by uttering 'She is hungry', demonstrating Barbarella. We will say that utterances which agree in import *samesay* each other. In L the samesaying relation is expressed by '$SS(u,u^\star)$'. So, Arabella utters:

(25) (Eu) $(U(\text{Barbarella},u)$ & $SS(u,\text{that}))$ [She is hungry.] [demonstrating Barbarella]

In asserting the first half (25), Arabella is asserting that Barbarella produced an utterance which samesays the utterance of the bracketed sentence in (25). Cinderella can use her knowledge of the truth conditions of each utterance which, together, compose an utterance of (25) to conclude, in accordance with patterns of inference now familiar, that:

(a) Barbarella made an utterance which samesays that, demonstrating the second part of (an utterance of) (25); and

(b) Barbarella is hungry.

The question naturally arises of when two utterances samesay each other. We can begin to answer it by considering the point of reports like (25). In uttering the demonstrated utterance in (25), Arabella is attempting to convey to Cinderella some of the information which would have been conveyed to her, had she heard and understood Barbarella's utterance. For Arabella's report to be correct, it need not convey all the information contained in Barbarella's utterance—that is probably impossible—but it is usually required that it convey the information that could be extracted in virtue of knowing the truth conditions of the utterance and certain of the features of contexts of the two utterances—for example the utterer, the time of utterance, the demonstrata, and so forth. In saying this we are not providing necessary and sufficient conditions for $SS(u,u^\star)$, but merely giving an initial explanation of the relation. We will suppose that the practices of indirect reporting in L suffice to fix (approximately) the extension of $SS(u,u^\star)$ for utterances of L. But it may prove impossible to explicate this relation in other terms, or even to provide an axiomatized theory of it. We will return to these points later.

Competent speakers of L know a correct truth theory for L, know that speakers of L know the truth theory, and grasp the samesaying relation sufficiently well to know generally whether $SS(u,u^\star)$. Is this knowledge sufficient to account for the fact that speakers of L understand L? We previously saw that knowledge of a truth theory (and even knowledge that the truth theory is common knowledge among a certain population) falls short of capturing one central aspect of linguistic competence. We saw that A's knowing a truth theory for L does not license inferences from 'B assertively uttered u' to 'B said that p', for example, from 'B assertively uttered 'Snow is white'' to 'B said that snow is white'. But speakers of L do possess information that licenses that inference from 'B uttered u' to '$(Eu)(U(B,u)$ & $SS(u,\text{that}))$'. For example, suppose A knows that B assertively uttered 'I am hungry'. Then knowing that this utterance samesays her own utterance (should she make one) of 'B is hungry' (or her thought of 'B is hungry') justifies her in concluding that $(Eu)(U(B,u)$ & $SS(u,\text{that}))$, where the demonstratum is an utterance (or thought) of 'B is hungry'.[4] If

(26) $(Eu)(U(\text{Barbarella},u)$ & $SS(u,\text{that}))$. [She is hungry.]

is a correct paraphrase of:

(27) Barbarella said that she is hungry,

---

[4] Arabella's thought that $(Eu)(U(\text{Barbarella},u)$ & $SS(u,\text{that}))$ [She is hungry.] will be true as well, and for the same reason, if we suppose that the thought itself has a paratactic structure; that is, it contains a demonstrative element referring to the thought that she is hungry. We are indebted to Stephen Schiffer and Ernest Sosa for helping us get clear about this.

then we have solved the problem of how to complete the inference from 'A assertively uttered S' to 'A said that *p*' within a truth-theoretic account of meaning.

The 'if' above is, of course, a big one. The way (26) functions in L is similar to the way (27) functions in English. However, there are numerous objections to the paratactic account—the claim that (26) is a correct paraphrase of (27)—that have been voiced. We turn next to a consideration of the most important of these objections.

2.

Semantic theory for some fragment of natural language has two aims. One is the construction of a truth theory for the fragment. The other is the systemization of logical properties of, and relations among, sentences of this fragment. These two aims are related and commonly thought to be simultaneously satisfiable by paraphrasing the sentences of the fragment into a formal notation, for which truth and truth-in-a-model are definable. Propositional attitude attributions, and specifically indirect discourse, have long been known to possess features which frustrate the construction of satisfactory theories of truth and implication. Perhaps the most discussed feature is the apparent failure of the principle of substitutivity of identity:

from $t = t\mathbb{C}$ and f(t), where 't' and '$t\mathbb{C}$' are singular terms and the wff f($t\mathbb{C}$) is obtained from the wff f (t) by replacing an occurrence of t by $t\mathbb{C}$.

Thus the inference from (28)–(29) to (30) is an apparent instance of substitutivity of identity, but it is invalid:

(28) Galileo said that the earth moves.
(29) The Earth = the third planet from the sun.
(30) So, Galileo said that the third planet from the sun moves.

One kind of response to this problem is to deny that the inference is really an instance of substitutivity of identity. Frege claims that the reference of a singular term depends on whether it occurs in a propositional attitude context (Frege, 1952, p. 57). In (29), 'the Earth' refers to the earth, but in (28) it refers to the sense of 'the Earth'. Quine's view is that singular terms in propositional attitude contexts do not refer at all. Thus 'the Earth' is no more a meaningful component of (28) than 'cat' is of 'concatenation' (Quine, 1960, Chapter 6). These accounts 'explain' the failure of substitutivity of identity by claiming that the semantic value of an expression in a propositional attitude context differs from its ordinary semantic value. The cost of this approach is a loss of *semantic innocence* (Davidson, 1968 p. 108). The semantically innocent believe that occurrences of the same word in and out of propositional attitude contexts possess the same meaning. Surrendering semantic innocence makes it difficult to understand how it is that one can learn from an utterance of 'Barbarella said that the cat sits behind the oven' that the cat sits behind the oven.[5]

One of the most striking features of the paratactic account is that, while it is similar to the Fregean and Quinean accounts in denying that the inference from (28)–(29) to (30) involves a genuine failure of substitutivity, it, unlike them, preserves semantic innocence. Explaining how this works will require us to add to our discussion above of the logic of first-order languages with demonstratives and indexicals. We will suppose that our language contains subscripted demonstratives: 'that$_1$', 'that$_2$', etc. The truth value of the utterance of a sentence with demonstratives and/or indexicals depends both on the sentence uttered and on the context of utterance. A context is specified by an n-tuple of features of a possible utterance or sequence of utterances, for example <a possible world, speaker of utterance, location of speaker, time of utterance, {a sequence of items to serve as demonstrata}, ... >. We assume that indexicals and demonstratives are logical constants constrained so that 'I' is always assigned to the speaker, 'that$_n$' is assigned to the nth demonstratum. We also assume that contexts are constrained so that the utterer is located at the place and time in the world of the context. 'S, therefore, S$\star$' is valid if and only if there is no context and no assignment of interpretations to the non-logical vocabulary in S and S$\star$, which makes S true and S$\star$ false. On this account, 'I am F and I am G, therefore, I am F' is valid, and so is 'I am here now'.

Let us see how this account applies to the paratactic paraphrases of indirect discourse sentences. According to the paratactic account, the first-order paraphrase of (28) is (31):

(28) Galileo said that the earth moves.
(31) (Eu)(U(Galileo,u) & SS(u,that)) [The earth moves.]

The paraphrase consists of two sentences. According to the account, a typical utterance of (28) is an assertion equivalent in import (and purport) to the first sentence in (31). Since (31) contains a demonstrative; it is not true or false except relative to a context. A typical utterance of (31) creates a context relative to which 'that' refers to an utterance of the bracketed sentence. It is obvious that, according to our account of validity, (31) implies 'Galileo said something', that is, '(Eu)(E$u\star$) (U(Galileo,u) & SS ($u,u\star$))', since, relative to any context in which (31) is true, 'that' refers to something which samesays Galileo's utterance (*pace* Arnauld, 1976, p. 289).

According to the paratactic account, the paraphrase of (28)–(30) is (31)–(33) respectively:

---

[5] Davidson's main objection to the Fregean and Quinean accounts is that it does not seem possible to construct finitely axiomatized truth theories for either of them, since they contain infinitely many primitive predicates. Davidson also mentions that these accounts are not semantically innocent. These two points are related, since it is the proliferation of words and their meanings that blocks finite axiomatization. But even if it were possible, as it seems to be for the Fregean theory, to produce a finitely axiomatized truth theory, the loss of semantic innocence would still be objectionable, since we would have no explanation of how someone is warranted in concluding that the cat sits behind the oven from an utterance of 'Barbarella said that the cat sits behind the oven.'

(31) (Eu)(U(Galileo,u) & SS(u,that)) [The earth moves.]
(32) The earth = the third planet from the sun.
(33) So, (Eu)(U(Galileo,u) & SS(u,that)) [The third planet from the sun moves.]

Since there are contexts in which the occurrences of 'that' in (31) and (33) refer to different individuals (strictly speaking, in L they must have different subscripts), this inference is invalid. It is also clear that the inference is not an instance of substitutivity of identity, since the singular terms 'the earth' and 'the third planet from the sun' do not even occur in (31) and (33). The expressions that occur in the demonstrated utterances possess their usual semantic values—their usual truth conditions—and so the account is semantically innocent. (It is just this feature that allows someone who knows the truth conditions for both 'Galileo said that' and 'the earth moves' to infer from an utterance of (31) and from Galileo's truthfulness that the earth moves.)

Despite these important virtues, the paratactic account has not won anything like general acceptance. The literature is replete with objections to the account. But, in view of how beautifully the paratactic account works in L to enable the speakers of L to fulfill the functions of indirect discourse, it seems to us to be worthwhile, though perhaps heroic and quixotic, to see to what extent the account, as an account of English indirect discourse, can be defended from these objections. Though it would be impossible to review and respond to every criticism in the literature, we have identified a few serious kinds of objections. Almost all the criticisms can be separated into three sorts:

(a) it is too strong;
(b) it is too weak; and
(c) it fails to generalize, for example, to other propositional attitudinal ascriptions and *de re* constructions.

In this chapter we address all these objections, with the exception of how to extend the account to *de re* attributions.[6]

It has seemed to a number of commentators that (28) is *about* Galileo and perhaps an utterance of his, but *not* about an English utterance, while its purported paraphrase (31) is about an English utterance. We doubt that intuitions concerning what a sentence is about should carry much weight by themselves in deciding the correctness of paraphrases. However, the objection would be a good one if it could be shown that, because of this difference, (31) implies or in some way requires the existence of something whose existence is not required by the truth of (28). The worry is that (31) implies or requires the existence of an English utterance, while (28) does not. Let us see if this is correct.

---

[6] A qualification on premise (59) is needed for utterances of sentences which do not contain context-sensitive features.

While it may seem that (31) implies 'there exist English utterances', in fact it does not. Given our characterization of implication for a language with demonstratives, (31) does not *imply* the existence of any utterance other than Galileo's. On that account, recall, S implies S⋆ if and only if there is no context and no interpretation of the non-logical vocabulary relative to which S is true and S⋆ is false. (31) is true relative to a context C if and only if there is an utterance made by Galileo which bears the samesay relation to the referent of 'that' in C. Consider a domain containing a single utterance $u^\star$ such that U(Galileo,$u^\star$) and SS($u^\star$,$u^\star$) and the context C relative to which 'that' refers to $u^\star$ Relative to that context, (31) is true and '(Eu) (Ex) (U(x,u) & u^1 $u^\star$)' is false. So (31) does not imply the existence of any utterance other than an utterance by Galileo.

Even though (31) does not imply 'there exist English utterances', still, according to the paratactic account, an utterance of (28) makes reference to an English utterance. Some authors think that a consequence of this is that the paratactic paraphrase of (34) is false, even though (34) is true (Lycan, 1973, p. 139; Blackburn, 1975, p. 184; Bigelow, 1980, p. 17). If this charge were correct, then it would seem to show that (28) and (31) are not equivalent. But the charge is incorrect. The paratactic paraphrase of (34) is (35):

(34) It is possible that Galileo said that the earth moves even though no English utterance ever existed.

(35) (Ex) (x = that) & P(U(Galileo,u) & SS(u,x) & (y) − English (y))) [The earth moves.]

In possible world semantics, (35) is true relative to a context in which the demonstratum of 'that' is an utterance $u^\star$ of the 'the earth moves' if and only if there is some possible world w and an utterance v made by Galileo in w and $u^\star$ samesays v in w and there are no English utterances in w. If the truth of SS(v,$u^\star$) in w required that $u^\star$ exist in w and be an English utterance, then (35) would be false. But neither of these seems to be the case. In particular, we see nothing wrong with supposing that SS(v,$u^\star$) may be true at w, even though v and $u^\star$ do not exist in the same world. Within the context of possible world semantics, it is not unusual to appeal to similarity relations across worlds—for example to the counterpart relation, which relates individuals in different worlds. Samesaying is such a relation.

There is another objection to the paratactic account; it is based on the fact that an utterance of (28) makes reference to an English utterance, and it is due to Brian Loar (reported in Schiffer, 1987a, pp. 131–133). Call an occurrence of a singular term in a sentence *primary* if and only if it is not properly contained within the occurrence of another singular term; for example, 'George' is not primary in 'George's car is red'. Loar claims that principle (P) is true:

(P) If the occurrence of t in 'A said that . . . t . . . ' is primary and refers to x, then that sentence is true only if A referred to x.

For example, suppose that Arabella utters 'the president is funny' and Barbarella reports her by saying 'Arabella said that Ronald Reagan is funny'. According to (P), Barbarella's report can be true only if Arabella referred to Ronald Reagan in making her utterance. If Barbarella said 'Arabella said that the star of *Bedtime for Bonzo* is funny', then her utterance still meets the test, since '*Bedtime for Bonzo*' has secondary occurrence.

Loar observes that the paratactic account results in a violation of (P) (and see also Burge, 1986, pp. 193–194; Blackburn, 1975, p. 184). According to (P), a problem is created for the paratactic account by (utterances of) sentences like (36):

(36) Laplace said that Galileo said that the earth moves.

An utterance of (36) may be true even though Laplace never referred to an utterance of English. But the paratactic paraphrase of (36) is (37):

(37) (Eu) $(U(\text{Laplace},u)$ & $SS(u,\text{that}))$ [Galileo said that the earth moves.]

According to (P), if (37) is a correct report, then Laplace must have referred to the referent of the second occurrence of 'that', since it has primary reference. But its referent is an English utterance of 'the earth moves', and it is certain that Laplace never referred to this utterance. Since (36) may be true even though Laplace never referred to an utterance of English, it follows, according to Loar, that the paratactic account must be mistaken. In short, Loar, noticing that on the paratactic account an utterance of (36) is about an English utterance, asks how Laplace could have said that, since he never referred to an English utterance. We are not impressed. (P) does not seem to us to be generally true, and in the case of iterated indirect discourse reports we can easily see how the paratactic account not only violates the principle but motivates its violation.

Here is an example of an apparent counterexample of (P). Barbarella utters 'I want a car like that', pointing at a Saab 900. Arabella correctly reports 'Barbarella said that she wants a car like that', pointing at a different Saab 900. Barbarella did not refer to the car which is the demonstratum of Arabella's second 'that'. A defender of (P) might counter that the occurrence of 'that' is really a secondary occurrence by claiming that the above sentence is really expressing 'I want a car of the same type as that'. 'The same type as that' has primary occurrence. But this paraphrase is itself controversial—it cannot be this easy to establish that types exist—and so it is unsuitable as part of a defense of a principle designed to establish the failure of the paratactic account. In any case, we can show how the paratactic account *motivates* a violation of (P).

Recall that, in L, the point of Arabella's indirectly quoting Barbarella to Cinderella was for Arabella to produce an utterance which would convey to Cinderella what Barbarella's utterance would have conveyed to her, had she been in a position to hear and understand it. Now suppose that Arabella, a speaker of L, attempts to conform to (P) when reporting what Laplace said. We assume that Laplace speaks a version of French for which the paratactic account is correct. Suppose that Laplace uttered 'Galilée dit que. La terre tourne'. If Arabella wants to report what Laplace said while conforming to (P), she needs to say 'Laplace said that. [Galileo said that]', where the demonstratum of the second 'that' must be the very same utterance Laplace demonstrated. Not only is this likely to be very inconvenient—Laplace's utterance is long gone—but, if her audience does not understand Laplace's utterances, that is, does not know their truth conditions, then the point of indirect discourse is lost. If instead Arabella demonstrates not Laplace's utterance but an utterance of her own, which samesays Laplace's utterance, then her audience (which we assume understands her utterances) will be in a position to learn that, if Galileo spoke the truth, the earth moves. If we are right, then Laplace's utterance of 'Galilée dit que la terre tourne' and Arabella's utterance of 'Galileo said that the earth moves' samesay each other even though they refer to different utterances. We see no problem with this. So, although it is true that on the paratactic account an utterance of (28) will be about an English utterance, we have seen that this entails neither that this sentence logically implies the existence of English utterances nor the falsity of the modal (34). And, while the paratactic account does lead to a violation of (P), that principle is not always true and, in the case of indirect discourse, it is justifiably violable.

A second kind of objection is that the paratactic paraphrase of an indirect discourse statement is logically weaker than the paraphrased statement (Arnauld, 1976, pp. 289–291; Platts, 1979, p. 123; Burge, 1986, p. 203; Schiffer, 1987a, pp. 134–137). For example, each of the following has been claimed to be logically valid:

(38) Galileo said that Osiander and Bellarmine are wrong.
(39) So, Galileo said that Bellarmine and Osiander are wrong.

(40) Galileo said that the earth moves.
(41) So, Galileo said that the earth moves.

(42) Galileo said that the earth moves.
(43) The earth moves.
(44) So, Galileo said something true.

(45) Galileo said that the earth moves.
(46) Everything Galileo said is true.
(47) So, the earth moves.

Each of these inferences appears to be valid, yet their paratactic paraphrases are not valid. The inference from (38) to (39) fails, since there are contexts relative to which the two occurrences of 'that' refer to different utterances. (The occurrences of 'that' will be paraphrased by 'that$_i$' and 'that$_k$', $i^1 \ne k$, since their actual demonstrata differ.) For exactly the same reason (40) does not imply (41).

Before concluding that this shows the paratactic account to be hopeless, let us note that this feature of the account is sometimes an advantage, especially when compared with competing accounts. According to these, 'A says that p' relates a person and a proposition—the proposition expressed by 'p'. These accounts differ in what they count as a proposition: a class of possible worlds, a Fregean thought, a structured entity, for example <individual, a property>. Consider each of the following inferences:

(48) Galileo said that the earth moves.
(49) So, Galileo said that the earth moves and (p or $-p$).

(50) Galileo said that bachelors are unmarried males.
(51) So, Galileo said that bachelors are bachelors.

(52) Galileo said that Cicero = Cicero.
(53) Galileo said that Cicero = Tully.

Each of the propositionalist accounts just mentioned validates one of these inferences, but each seems invalid. Defenders of propositionalist accounts have devoted a considerable amount of effort attempting to persuade their readers (and themselves?) that the inference each one claims to be valid really is valid (Hintikka, 1969; Stalnaker, 1984; Salmon, 1986). But their defenses seem to us to be lame. It is difficult to believe, for example, despite her protestations to the contrary, that Lois Lane believes that Superman is Clark Kent (see Schiffer, 1987b). The paratactic paraphrases of each of the above inferences are, of course, invalid for exactly the same reason why the inference from (38)–(39) is invalid. But pointing to the faults of other accounts does not show that the paratactic account is correct about the inferences with which we began this discussion.

We begin by talking about the paratactic paraphrase of the arguments (42)–(44) and (54)–(56) respectively (see Davidson, 1969 pp. 50–52):

(54) $(Eu)(U(\text{Galileo},u)$ & $SS\ (u,\text{that}_1))$ [The earth moves.]
(55) The earth moves.
(56) So, $(Eu)(U(\text{Galileo},u)$ & $\text{true}(u))$

It is obvious that (54)–(56) is invalid. However, if we add to this argument premise (57), the argument is valid:

(57) That$_1$ is true iff the earth moves.

The additional premise (57) expresses a truth known to anyone who knows that 'that$_1$' refers to an inscription of 'the earth moves' and who understands this inscription. Of course, any English speaker who examines this argument will know this. So the paratactic account has a ready explanation for why the argument (54)–(56) seems to be valid to an English speaker even though it is an enthymeme. A perfectly similar account can be given of the argument (42)–(44).

An explanation can also be given of the apparent validity—apparently, that is, according to the paratactic account—of the inferences (38)–(39) and (40)–(41).

The missing premises in the first of these inferences are:

(58) SS(that$_1$,that$_2$)

(59) (u) $(u\cent)$ $((\text{True}(u)$ & $SS(u,u\cent)) \rightarrow SS(u,u\cent))$[7]

Premise (58) will be known by anyone who knows that that$_1$ is of 'Osiander and Bellarmine are wrong' and that that$_2$ is of 'Bellarmine and Osiander are wrong' and who understands English. Recall that someone who understands English is able to recognize when an utterance of a sentence samesays an utterance of another sentence. So, according to the paratactic account, these two inferences are not valid, but they can be turned into valid inferences by adding premises known to any English speaker who attends to them.

The fact that the inferences (42)–(44) and (45)–(47) are not valid according to the paratactic account suggests a slightly different objection. As we have seen, on the paratactic account it is logically possible, for example, for someone to know that Galileo said that the earth moves without understanding what Galileo said. This follows from the fact that, on the truth-theoretic account of meaning, what Galileo said includes knowing the truth conditions of what Galileo said. The fact that speakers of English who hear someone utter 'Galileo said that the earth moves' and believe the utterance to be true will understand what Galileo said only partially answers this objection. It must be admitted that, on the paratactic account, A may know that B said that *p* without understanding what B said. But is this so obviously a bad consequence? Think of A, who mistakenly enters a room in which a physicist is lecturing and hears the lecturer utter 'we conducted three tests of Bell's inequalities'. B asks A what the physicist said, and A replies: 'he said that he and his colleagues conducted three tests of Bell's inequalities – but I must admit I did not understand what he said since I have no idea what Bell's inequalities are'. There seems to us to be nothing preventing all of A's utterances from being true, contrary to the assumption that underlies the objection.[8]

A third kind of objection to the paratactic account concerns whether it can be extended to propositional attitude reports other than 'says that', to *de re* attributions, and to interrogative constructions ('A said who the Prime Minister is'). There are formidable problems to be confronted in developing such extensions and we have by no means solved them all. However, we do think that the situation is considerably more promising than is commonly thought. Here we will confine ourselves to showing how the paratactic account can be extended to propositional attitude reports.

The natural extension of the paratactic account to belief reports is to paraphrase, for example, (60) as (61):

---

[7] The intelligibility of A's remarks suggests that the relation between knowing what someone said and understanding what he said is one of conversational implicature.

[8] This relation may have a functional characterization and the state may possess syntactic properties, but we are not assuming either.

(60) Galileo believes that the earth moves.
(61) Galileo believes that. [The earth moves.]

According to the account, 'believes that' relates a person to an utterance, so that (61) is true just in case Galileo bears this relation to the demonstrated utterance of 'the earth moves'. Schiffer thinks that this account cannot be correct (see also Haack, 1971, pp. 360–61; Leeds, 1979, p. 51; Bigelow, 1980, p. 17). Schiffer writes:

> The representation of the saying—that relation as ['$(Eu)(U(Galileo,u)$ & $SS(u,that)$)'] is plausible because if 'Galileo said something' is true, then there can be no barrier to inferring '$(Eu)S(Galileo, u)$' for there is always Galileo's own utterance to be an utterance to which he stands in the saying-relation as portrayed in ['$(Eu)(U(Galileo,u)$ & $SS(u,that)$)']. But if 'Galileo believed something' is true, then there *is* a barrier to inferring '$(Eu)B(Galileo,u)$;' namely, that there may not be any actual utterance that gives the content of Galileo's belief. [. . .] Believing could be represented as a relation to actual utterances only if one could be assured that for every belief there was some actual utterance that gave the content of that belief; but of course one cannot be so assured. (1987a, pp. 126–127)

Schiffer thinks that the only revision open to the paratactic account is to declare that belief relates a person, not to an utterance, but to an utterance-kind, and this revision undermines Davidson's extensionalism. But Schiffer is mistaken. There is a viable alternative.

It is relatively uncontroversial that someone believes that p just in case he is in a token (brain) state which has the content that p. Let $B(a,s)$ be the relation that holds between an individual a and a token state, for example an event of neuron firings, which is a belief state of a. The full paratactic account of (60) is (62), and (63) is paraphrased by (64):

(62) (Es) $(B(Galileo,s)$ & $SS(s,that))$ [The earth moves.]
(63) Galileo believes something.
(64) $(Es)B(Galileo,s)$

(64) does not entail that there exists any actual utterance which samesays s. Our account assumes that belief states and utterances can samesay each other. We think this is untendentious.

It seems possible to extend this account to other kinds of propositional attitudes. (65) is paraphrased as (66):

(65) Galileo desires that the earth moves.
(66) $(Es)(D(Galileo,s)$ & $SS(s,that))$ [The earth moves.],

where $D(a,s)$ holds of a person and a token state of desire. Sentences like (67) seem to create problems for this approach (Higginbotham, 1986, p. 39), since they do not contain 'that' and the complement of 'wants' is not a sentence. But we tentatively suggest that (67) be paraphrased by (68):

(67) Galileo wants the earth to move.

(68) $(Es)(W(Galileo,s)$ & $SS(u,that))$ [The earth to move.]

This paraphrase assumes that the content of Galileo's mental state is the same as the content of the demonstrated utterance of 'the earth to move'.

Of course, even if it is granted that our replies to the various objections help to deflect them, one still might wonder why it would not be better to have an account which completely avoids them. That is, wouldn't it be better to have an account which entails that Galileo's utterance is true if and only if the earth moves, and, given that it is true that the earth moves and that Galileo said that the earth moves, then Galileo said something true *without the help of all the additional premises*, and something which did not have utterances of 'Galileo said that the earth moves' being about an English utterance, and so on? And in fact, such theories are readily available—namely, the Fregean theory, for example, which claims that sentences in indirect discourse (or other propositional attitude sentences) express a relation between a person and (not an utterance but) a *proposition*. On such an account, (28) is true if and only if (69):

(28) Galileo said that the earth moves.

(69) (Ep) $(p =$ the proposition that the earth moves and Galileo says p).

On Frege's view, the words on the right of 'that' refer to their usual senses. So the sentence says something like this: Galileo said the thought composed of such and such senses in such and such ways. The reasons why propositions seem so well suited to be the relata of propositional attitudes are, first, that they are abstract entities, so that (69) does not entail the existence of any utterance of an utterer of (28), and, second, that they have the truth conditions essentially, so that (69), together with 'it is true that the earth moves', entails that 'Galileo said something true'.

We have shown that none of the objections to the paratactic account we have considered is fatal to it. The account, like other accounts (Russell's theory of descriptions, possible world accounts of subjunctive conditionals, Davidson's account of action and event sentences), has consequences which seem surprising to our intuitions. But we think we have shown that the account is sufficiently explanatory of our practices of indirect discourse and attitude reports to be counted as plausible, more plausible than its propositionalist rivals. If the paratactic account is correct, then, as we have shown, it provides a way of overcoming what has been a vexing problem for truth-theoretic accounts of meaning: the problem of specifying what someone says in uttering u.

Acknowledgments

Earlier drafts of this chapter were read at the Universities of Bologna, Milan, Minnesota, Siena, Modena, Pisa, Venice, and at the Conference on Information Based Semantic Theories in Tepotzlan, Mexico. We would like to thank all those who have helped us improve this chapter, in particular John Biro, Andrea Bonomi, Michael

Hand, James Higginbotham, Paolo Leonardi, Ernesto Napoli, Eva Picardi, Stephen Schiffer, and the graduate students in Lepore's philosophy of language seminar, spring 1988.

References

Arnauld, R. (1976) Sentences, Utterance, and Samesayer. *Nous* 10: 283–304.

Bigelow, J. (1980) Believing in Sentences. *Australasian Journal of Philosophy* 58: 11–18.

Blackburn, S. (1975) The Identity of Propositions. In S. Blackburn (ed.), *Meaning, Reference, and Necessity*, Cambridge: Cambridge University Press, pp. 182–205.

Burge, T. (1986) On Davidson's 'Saying That'. In Ernest Lepore (ed.), *Truth and Interpretation*, Oxford: Basil Blackwell pp. 190–210.

Davidson, D. (1967) 'Truth and Meaning', *Synthese*, 17: 304–323; reprinted in Davidson (2001).

Davidson, D. (1968) 'On Saying That', *Synthese*, 19: 130–146; reprinted in Davidson (2001).

Davidson, D. (1969) 'True to the Facts', *Journal of Philosophy*, 66: 748–764; reprinted in Davidson (2001).

Frege, G. (1952) 'On Sense and Reference' [1892]. In P. T. Geach and M. Black (eds), *Translations from the Philosophical Writings of Gottlob Frege*, Oxford: Blackwell Publishers, pp. 56–78.

Haack, R. J. (1971) On Davidson's Paratactic Theory of Oblique Contexts. *Nous* 5: 351–361.

Higginbotham, J. (1986) Davidson's Program in Semantics. In Ernest Lepore (ed.), *Truth and Interpretation*, Oxford: Basil Blackwell, pp. 29–48.

Hintikka, J. (1969) Semantics for Propositional Attitudes. In J. Davis, D. Hockney, W. Wilson, (eds), *Philosophical Logic*, Dordrecht: D. Reidel, pp. 21–45.

Leeds, S. (1979) Church's Translation Argument. *Canadian Journal of Philosophy* 9: 43–51.

Lycan, W. (1973) Davidson on Saying That. *Analysis* 33: 138–139.

Platts, M. (1979) *Ways of Meaning.*, London: Routledge and Kegan Paul.

Quine, W. V. O. (1960) *Word and Object*. Cambridge, MA: MIT Press.

Salmon, N. (1986) *Frege's Puzzle*. Cambridge, MA: MIT Press.

Schiffer, S. (1987a) *Remnants of Meaning*, Cambridge, MA: MIT Press.

Schiffer, S. (1987b) The 'Fido'–Fido Theory of Belief. In James Tomberlin (ed.), *Philosophical Perspectives I*, Atascadero, California: Ridgeview Publishing Company, pp. 455–480.

Stalnaker, R. (1984) *Inquiry*. Cambridge, MA: MIT Press.

8

Conditions on Understanding Language

Ernest Lepore

1.

Philosophers in general are uncomfortable, if not downright skeptical, about attributing semantic knowledge, particularly of a semantic *theory*, to ordinary speakers (see for example Quine, 1972; Thomason, 1974, Hacking, 1975; Dummett, 1974; Foster, 1975; Devitt, 1981; Evans, 1981; Baker and Hacker 1984, Schiffer, 1987; Wright, 1986.). Those who do not feel the pinch often adopt a two-pronged defence: they rebut skeptics with an array of distinctions (and hedges), contending that the skeptics' confusions arise because they ignore such distinctions (Davies, 1981; Lepore, 1982; Higginbotham, 1983, 1987; George, 1989), and at the same time they argue that attributing such knowledge provides the best working available *empirical* hypothesis to account for linguistic comprehension (Partee, 1975, 1979; Evans, 1981; Higginbotham, 1988, 1995; Larson and Segal, 1995; Segal, 1994). Though sceptical arguments about the relevance of semantics in explicating linguistic comprehension abound (Chomsky, 1986; Fodor, 1975, 1987; Hornstein, 1984; Stich, 1983; Schiffer, 1987; Soames, 1989), a more acute challenge issues from Fodor and Schiffer: each offers an account of language understanding that *excludes* metalinguistic (semantic) knowledge, and therefore knowledge of semantic *theory* (Fodor, 1984, 1989; Schiffer 1987, 1994; see also Harman, 1975).

Both Schiffer and Fodor deny that there is any more to mastering a language than coming to have a capacity to go from what is heard to what is said. Schiffer states

A theory of understanding for a language L would explain how one could have an auditory perception of the utterance of a novel sentence of L and know what was said in the utterance of that sentence. (Schiffer, 1987, p. 113; see also p. 262).

And Fodor continues

I assume that language perception is constituted by non-demonstrative inferences from representations of certain effects of the speaker's behavior (sounds that he produces, marks that he makes) to representations of certain of his intentional states, [in particular] a canonical representation of what the speaker said. (Fodor, 1984, pp. 5–6)

Fodor and Schiffer deny that knowledge (or any other kind of epistemic/doxastic/psychological attitude) about the semantics of one's spoken language is causally relevant for effecting these transitions.

[W]hen we understand the utterance of a sentence we do not first *come to the belief* that it means such-and-such and then have that as our basis for thinking that the utterer was saying such-and-such. (Schiffer, 1987, p. 262, my emphasis)

What really matters is this: For any perceptually analyzable linguistic token there is a canonical description (DT) such that for some mental state there is a canonical description (DM) such that 'DTs cause DMs' is true and counterfactual supporting. (Fodor, 1989, p. 8)

On this view, understanding consists in being a template for a causal network between the perceived linguistic sounds and shapes on the one hand, and subsequent internal mental states on the other. Understanding, adverting to jargon, consists in having a linguistic module in the mind. This module works (however it may) to set up a certain law-like co-variation between instantiations of categories in the world and concepts in the mind:

The translation algorithm [from English into Mentalese] might well consist of operations that deliver Mentalese expressions under syntactic descriptions as output given English expressions under syntactic descriptions as input with *no semantics* coming in anywhere except, of course, that if it's a good translation, then semantic properties will be preserved. (Fodor 1990, pp. 187–8, my emphasis)

Schiffer concurs: no internally represented semantics is required for the use of a public language *even if* it has a semantics (Schiffer, 1987, p. 116). This is the lesson of Schiffer's Harvey counter-example (ibid., pp. 192–207):

Harvey thinks in Mentalese [. . .] and his language processing uses not an internally represented [meaning theory] of English but rather an internally represented *translation manual* from English to Mentalese [. . .] Such a theory assigns no semantic values to the expressions of either language and in no sense determines a grammar (i.e., a meaning theory) for either language. (Schiffer, 1987, pp. 192f; see also p. 262 and Schiffer, 1994, p. 304)

Since Schiffer contests the need for an internally represented compositional semantics, the question whether *any* epistemic relationship between a speaker and this semantics need exist cannot even arise. So, if the Fodor/Schiffer account—*translationism*, merely to label it[1]—is correct, then the semanticist's is not, as Fodor likes to say, the only game in town.[2]

But how could translationism be right? Isn't it obvious that speakers of Italian, for example, know semantic facts like (1) below:

(1) 'Sta nevicando' means that it's snowing.

---

[1] Actually 'transductionism' might be a better label, since it suggests an analogy with perception and since the notion of translation that Schiffer and Fodor are employing bears little resemblance to what we ordinarily call 'translation' (see Fodor, 1984).

[2] Hornstein (1984), travelling on a different route, arrives at a position similar to translationism.

Translationists *can* agree that (1) is a truth which Italian speakers know. But why, they wonder, *must* such knowledge be invoked in order to understand Italian? (1) is true only because speakers of Italian use 'sta nevicando' to communicate ('encode' or 'express') the thought that it's snowing (Schiffer, 1993; 1994, pp. 303–304).[3] So, if someone knows that (1) is the case, this must be because he knows that his words can express this thought. From this it does *not* follow that such knowledge is (or need be) utilized in understanding 'sta nevicando'. This is *all* that Fodor means when he writes that he is 'Gricean in spirit though certainly not in detail' (Fodor, 1975, pp. 103–4; see also Fodor, 1987, p. 50 and Schiffer, 1982, p. 120, 1994, p. 323). Fodor—and even Schiffer now—are Gricean *only* inasmuch as both hold that, whatever semantic properties natural language expressions have, they are inherented from (a language of) thought, Mentalese. So, contra Dummett (1978, p. 97), translationists maintain that semantic knowledge about language is *inconsequential*; it plays no causal (and therefore no rationalizing) role in linguistic competence.[4]

Still, aren't translationists postponing the inevitable? Suppose, as Harman believes, that English speakers think 'in English'. Wouldn't it follow that understanding requires invoking knowledge about the semantics for English expressions? It would not. Even if one's *lingua mentis* is one's own public language, *pace* Fodor (1975, p. 79ff) and Schiffer (1987, p. 187), understanding still involves only the capacity to make transitions from what is heard to what is said, regardless of which language we think in. As Harman (1975) puts it, 'words are used to communicate thoughts that would ordinarily be thought in those or similar words' (p. 271). So, if we think 'in English', then, when someone who understands English hears an individual A utter 'it's snowing', there will go, into her 'belief-box', an English sentence to the effect 'A said that it's snowing'.[5] No assumptions about the semantic properties of A's words need be invoked, much less knowledge about these properties.

But, one might wonder, how can merely tokening an English *sentence* in one's belief box suffice for understanding? Mustn't we understand these internalized English

---

[3] I am going to circumvent the debate between Dummett (1993) and McDowell (1988) about the relationship between thought and language, namely whether language is a code or an expression of thought. As far as I can see, even if it were true that there is no having thoughts without having the words to express them, nothing follows about the causal efficacy of semantic knowledge in linguistic comprehension.

[4] I also won't address what has become a rather large rift between Fodor and Schiffer, namely whether there must be a compositional semantic theory for Mentalese itself. Fodor argues that thought is productive and systematic because Mentalese is productive and systematic, and this is best accounted for by assuming that there is an internalized compositional semantic theory for Mentalese (Fodor, 1987, 1991). Schiffer, on the contrary, argues, first, that any language rich enough to express propositional attitudes lacks a compositional semantic theory—and so Mentalese lacks a compositional semantic theory—and, second, that he can show how productivity and systematicity could be otherwise explained (see Schiffer, 1987). These issues, although important, are irrelevant to the topic of this chapter.

[5] *Pace* Segal (1994, p. 115), there is no need to view this function as mapping structural descriptions of representations of natural language expressions into structural descriptions of representations of a language of thought (see n. 2 above). If I am right, then there is no way for Segal to run a translation-style argument à la Davidson against translationism (see below).

sentences as well? Harman replies that we don't understand thoughts; we merely have or entertain them. Schiffer (1994) puts the point this way: 'understanding a [language of] thought is simply a matter of thinking in it' (p. 322). We don't say that he is thinking it's raining but doesn't understand his thought. This could only mean he is unclear in his *conception* of rain, or some such thing. So nothing *need* be known in order to understand thought. Translationism, in short, maintains:

[I] Understanding a natural language involves nothing more than having a (per-ceptual) capacity to hook up the natural language expressions with symbols of (the language of) thought.

[II] This skill requires no metalinguistic (semantic) knowledge.

My aim in this chapter is to refute [I] and [II].[6] Before embarking, I'd like to say more than a few words about how the challenge of translationism illuminates anew what most philosophers argued was a misguided research programme.

Structural semantics: a misguided research programme?

In the 1960s and early 1970s, Fodor (along with Katz and Postal) practised structural semantics (hereafter, SS). SS theorists countenance properties and relations like synon-ymy, antonymy, meaningfulness, anomaly, logical entailment and equivalence, redun-dancy, and ambiguity as a good initial conception of the range of semantics. Shunning details, SS theorists proceed by translating (or mapping) natural language expressions into (sequences or a set of) expressions of another language. There is no uniformity among them about the nature of this language or about how these translations or mappings are to be effected, but I beg no interesting questions by restricting attention to Katz and Fodor's SS proposal (1963). The culmination of the various mapping rules and other apparatus within the Katz/Fodor framework results in theorems like (2):

(2) 'Sta nevicando' in Italian translates (or is mapped) into the language Semantic Markerese as S.

Mappings like (2) are constrained, and this is the *raison d'être* for them as well as for semantic markers, such that synonymous expressions of a language L translate into the same (sequence or set of) expressions of Semantic Markerese, ambiguous expressions of L translate into different expressions of Semantic Markerese, anomalous expressions of L translate into no expression of Semantic Markerese at all and so on.

---

[6]   There's a lot to fuss about here. For example, Fodor and Schiffer are committed to the view that thought is (logically) prior to public language. So I could write a paper detailing why this language-of-thought hypothesis is wrong, but I won't. Besides, I agree with Fodor and with Loewer and Rey that the supposition that there is a language of thought, Mentalese, 'is best viewed as simply the claim that the brain has logically structured, causally efficacious states' (Loewer and Rey, 1992, p. xxxiii).

Once they surmised what SS was about, dissenting semanticists could not get into print fast enough to explain to Fodor and Katz just how confused they were about semantics. Davidson (1967, 1973), Vermazen (1967), Lewis (1970), Cresswell (1978), and Partee (1975) (among many others) argued that, since SS theories do not articulate relations between expressions and the world, they cannot provide an account of the truth conditions for such sentences, and therefore SS theories are not really semantic. Critics charged that the phenomena SS concerns itself with represent only a small portion of the full domain of semantics, and SS, they argued, cannot accommodate this full domain. As David Lewis put it: 'we can know the Markerese translation of an English sentence without knowing the first thing about the meaning of the English sentence; namely, the conditions under which it would be true. Semantics with no truth conditions is no semantics' (1970, pp. 169–170). Critics protested that, even if an SS for Italian assigned an interpretation to every Italian sentence, it would not specify what any expression of Italian means. On a Davidsonian conception, an adequate semantic theory for L must not only ascribe meanings to expressions of L, it must also ascribe them in a way that enables someone who knows the theory to understand these expressions. SS fails on this account, because in the overall picture of SS there are three languages: the natural language; the language of the semantic markers; and the translating (or mapping) language, which may be Semantic Markerese, the natural language, or some other language (Davidson, 1973, p. 129). Since SS proceeds by correlating the first two languages using the third, one can understand its mappings, for example (2), knowing only the translating (or mapping) language and not the other two. We can know that (2) is the case, perhaps on the basis of what Katz and Fodor tell us, without knowing what either 'Sta nevicando' or its Semantic Markerese translation S means.

If someone understands Semantic Markerese, he can no doubt utilize (2) to understand the Italian sentence; but this is because he brings to bear two things which he knows that (2) does not state: that Semantic Markerese is a language he understands; and whatever information he has in virtue of which he understands S. It is this latter information that an adequate semantics must characterize.

Fodor's revenge

Shortly after the assault on SS by practically the entire philosophical community, Fodor ceased doing semantics. Most philosophers thought (certainly I did) that he had acceded to his critics. However, in 1975, Fodor wrote that he saw little difference, if any, between specifying meaning by translation and specifying it with truth conditions:

We're all in Sweeney's boat; we've all gotta use words when we talk. Since words are not, as it were, self-illuminating like globes on a Christmas tree, there is no way in which a semantic theory can guarantee that a given individual will find its formulas intelligible [. . .] So the sense in which we can 'know the Markerese translation of an English sentence without knowing [...] the conditions under which it would be true' is pretty uninteresting. (Fodor, 1975, pp. 120–121)

In a critical response, Barry Loewer and myself argued that Fodor misunderstood either Lewis' objection against SS or the nature of truth-conditional semantics (see Chapter 1 of the present volume). I took Fodor to be misconstruing Lewis as saying that one must understand the language in which the canonical representation is expressed before one can utilize a semantic theory to determine what the represented sentences means; and that's a problem every semanticist faces. This certainly is correct, but Lewis' point is not this obvious one. Instead, he is arguing that someone who understands a translation and knows it to be true *need not* understand the sentence of the translated language. We cannot understand (3) or (1) unless we understand English. But knowledge that (1), unlike knowledge that (3), requires no familiarity with English. Simply note that, whereas (1) and (3) are grammatical, (4) is not:

(3) 'Sta nevicando' in Italian translates 'it is snowing' in English.
(4⋆) 'Sta nevicando' in Italian means that it is snowing *in English*.

One need know no more English to know that (1) is the case than Galileo knew for us truthfully to say that he said the earth moves.

With respect to the question whether language mastery requires semantic knowledge, none of this shows any more than that a semantics that specifies truth conditions for sentences of a language L may serve to characterize (at least partially) knowledge sufficient for understanding L, while a semantics that specifies translation from L into another language by mapping structural descriptions of L into structural descriptions of the latter cannot (unless we add the assumption that the latter language is known). The conjectural inference from 'knowledge of truth conditions (partly) *suffices* to understand L' to 'knowledge of truth conditions (partly) *constitutes* this understanding' may seem natural, perhaps even good science, but Fodor, for one, balks. What's wrong with translationism, he asks?

Fodor's current skepticism about the utility of semantics for natural languages to account for linguistic comprehension is more than congenial given his early commitment to SS. He has repudiated much of the original SS program: its commitment to an analytic/synthetic distinction (Fodor and Lepore, 1992); its commitment to certain views about lexical decomposition (Fodor, Fodor, and Garrett, 1975; Fodor, Garrett, Walker, and Parkes, 1980). But his early commitment to SS is of a piece with his denying the cogency of semantics for natural languages, at least qua theory of understanding.

Once Fodor gave up on the idea that understanding requires knowledge of a semantic theory, he began to see the semanticist's emphasis on natural language as sweeping under the rug all the interesting philosophical questions about content; for example, what bestows intentional (that is, contentful) states on a cognitive system. Semantic theories, whether of the SS or truth-conditional variety, are useless here.

Since natural languages, according to Fodor, merely *shadow* real intentionality, the (philosophical) explication of semantic properties must focus on the mind—in fact, on the semantic properties of symbols in the mind. But explicating the semantic properties of thought, according to Fodor, is a metaphysical enterprise. Metaphysical questions merit metaphysical answers; not epistemic or psychological ones.

How does this make Fodor unrepentant? A truth-conditional account is *no better* than the original SS translation account not because both employ language, but rather *because* both leave unanswered interesting metaphysical questions about intentionality. Philosophers, qua semanticists for natural language, need not, nor should they be expected to, answer these questions. Fodor concurs. He concludes that semantics for natural languages, worse than boring, is worthless. This is his real challenge. Translationists challenge semanticists to supply a purpose for their endeavour. The rest of this paper assumes this challenge by advancing considerations that incline me to conclude that:

(a) Understanding a natural language requires metalinguistic knowledge.
(b) This metalinguistic knowledge must be semantic, and so cannot be merely translational.

Understanding and rationalization

Maria utters to Massimo 'sta nevicando'. Because translationism is *compatible* with Massimo believing that Maria said that it's snowing without his believing anything about the (causal) connection between what he heard and what he believes she said, translationism seems equally compatible with Massimo believing that Maria said that it's snowing that Massimo foster only *false* beliefs about what Maria's words mean. But how, then, can the capacity to make correct transitions from what is heard to what is said alone suffice for linguistic competence? How can someone who has *only* false beliefs about what the expressions of a language L mean be linguistically competent with L? At least one author seems to think that it's possible. Mark Richard writes that someone 'might hold a false theory about competence, but still himself count as competent' (1992, p. 45). If 'false theory' means false beliefs about what words mean, then I disagree—unless by 'count' Richard means that the individual might never be exposed (something of no philosophical significance). So is semantic scepticism compatible with translationism?

Translationists, because going Gricean is an option, can say it is not. By virtue of being linguistically competent in their sense and by virtue of knowing what thoughts words express or encode, translationism can frustrate semantic scepticism. But this concedes nothing to those who insist that semantic knowledge is partly *constitutive* of linguistic competence, since semantic knowledge, on this account, is of no *consequence*. So, even if it were *impossible* to be linguistically competent without having true beliefs about what one's words mean, nothing follows about the causal efficacy of these beliefs in rendering the transitions translationists identify as constitutive of understanding. Even if intuition makes one incline toward authority about what one's words mean (perhaps because we have authority about what our thoughts are, according to translationism), no one is inclined to endow speakers with authority about the causal ancestry of their beliefs, in particular their beliefs about what another says when he speaks (or even with authority about the causal history of beliefs concerning what their words mean). This, unfortunately, is really bad news for philosophers/cognitive scientists who hope to frustrate translationism on empirical grounds (for instance Segal, 1994, pp. 116–117). Even if the best psychological account available of linguistic comprehension attributes rich semantic knowledge to competent speakers, this cannot establish that translationism is false, since it, too, can attribute such knowledge. To defeat translationism, we must establish that semantic information has *repercussions* for understanding. Neither the impossibility of semantic skepticism nor empirical science can establish anything so strong. I will take a different tack.

What I want to argue is that translationism is *inconsistent* with Massimo having *reasons for* his new belief. Like Dummett, I want to maintain that 'any adequate account of language must describe it as a *rational* activity' (1978, p. 104, my emphasis). Though Dummett's target is 'a causal theory such as Quine appears to envisage, representing [language mastery] as a complex of conditioned responses' (ibid.), I want to cast my net wider, so as to encompass translationism—a position, unlike Quine's, which is thoroughly cognitive. To this end I must show that the rationalizations that speakers qua speakers have *cannot* be underwritten by translationism.

Why should the sort of rationalizing which linguistic comprehension carries require ascribing metalinguistic (semantic) knowledge to speakers? Why isn't it secured already by the *reliable* connections between heard utterances and internal states of the mind, as translationism presumes? The capacity for language comprehension produces correct internal states on the basis of what is heard; so, why aren't such states justified on this basis alone?

Someone's belief being justified and that person's having another belief, which rationalizes his belief, are distinct. Many perceptual beliefs are justified directly by experiences on which they are based, and in principle a belief can be justified simply by being the result of a reliable belief-forming mechanism. One's belief that one is currently in pain is clearly not justified on the basis of other beliefs one has. The only explanatory story we are in a position to give is one which invokes a mechanism that connects reliably one's being in pain with one's believing that one is in pain. Translationists see linguistic comprehension in the same light (Fodor, 1984).[7]

What's on offer is reliabilism (Dretske, 1981; Goldman, 1986; Nozick 1981). Beliefs about what is said count as justified just in case processes that produce them tend, in the 'relevant' set of counterfactuals, to be truth-inducing. There is indeed a lawlike

---

[7] Of course, nothing comes easy. If perception requires propositional knowledge, not any appeal to an analogy with perception will bolster translationism (see Fodor, 1989, pp. 9ff; Fodor, 1984; and Bruner, 1957).

correlation between an Italian speaker's beliefs about what is said and the heard utterances that bring them about.

Also note that views of this sort do not require a KK principle, according to which being justified that p entails being justified that you are justified that p. This reliabilist feature serves translationism well. If someone's belief fixation processes may be reliable and constitute justification even though he does not realize that they do, then translationists can deny that speakers require special metalinguistic knowledge about the connection between what is uttered and what is said in order to secure whatever justification linguistic comprehension requires.

Much has been written for and against reliabilism, but in this context its viability is not relevant. We want to know Massimo's reason for his belief that Maria said that it's snowing when he heard her utter 'sta nevicando'. That he has a certain faculty that, *ceteris paribus*, delivers him from heard Italian utterances to true beliefs about what is said fails to reveal his reason. If Massimo knew he was so constituted that, by virtue of learning Italian, he reliably acquires true beliefs about what is said when he hears Italian utterances, then Massimo would have a reason for his belief. But drawing on such knowledge here is illegitimate. It undercuts reliabilism's appeal by resurrecting the KK principle.[8]

So, does Massimo have a reason for his belief *even if* he lacks beliefs about Italian? Imagine, as is consistent with translationism, that Massimo is clueless about why he believes (correctly, let's suppose) that p when he hears Maria utter something Italian. Nothing *in his head* justifies his belief. Massimo's condition is mildly pathological. Poor dupe, running around the world telling everyone he meets what others said, but always lacking reasons for such attributions. Massimo cannot explain his belief that Maria said that it's snowing any better than by saying: 'I don't know why I believe this. I just do. Didn't Maria say sta nevicando?' But someone who understood not one word of Italian and happened to find himself believing that Maria said that it's raining could make the same case for himself. Massimo is not unlike someone who perpetuates ghastly deeds, but literally has (or, should I say, *can* have) no idea why he persists. No degree of prodding or assistance could bring him to reconstruct reasons for his behaviour. Just as we would withhold agency from him, we should withhold linguistic comprehension from clueless Massimo.

Diagnosis: even if reliabilism secures some sort of justification, it does not secure one that is sufficient to underwrite linguistic comprehension. If Massimo believes that

---

[8] Reliabilism is not committed to the view that we can *never* know that we know certain things, but rather to the denial of the view that this second-order knowledge is always required. So, if Massimo knows he's reliable, why doesn't this provide a robust reason for his belief that Maria said it's snowing when he heard her utter 'sta nevicando'? Massimo's reason, then, is that he knows he's a reliable disquoter of Italian utterances. His justification is, I suppose, something like this: people tell him he's very good; he has no trouble communicating with Italian speakers, and so on. Despite any apparent initial plausibility, this view is absurd. I do not come to believe I understand my first language because I see others treat me as though I'm correctly disquoting them. Any such discovery comes long after I'm proficient in my mother tongue.

Maria said it's snowing when he hears her utter 'sta nevicando', we expect him to have beliefs about Maria's utterance that *play* a *rationalizing* role. If such a rationalization is integral to language understanding, where could it spring from, if not from knowledge (or belief, or other propositional attitudes) about the sounds and shapes of the language itself?[9]

Massimo, in our imagined scenario, really is clueless. He has no idea why he believes Maria to have said that it's snowing when he heard her utter 'sta nevicando'. So, even though he makes the transition, he has no reason—*conscious, unconscious, tacit, explicit, implicit,* or of any other sort. I am claiming that, if one understands a language, one must have reasons that rationalize her transitions. To echo Davidson, 'nothing can count as a *reason* for holding a belief except another belief' (1986, p. 123, my emphasis). I know no argument that defends this position *tout court*, but it seems right in the case of language comprehension. What about linguistic comprehension providing reason for the belief that it's snowing, when this understanding combines with the belief that Maria uttered 'Sta nevicando'? Additional beliefs or knowledge, which Massimo has about Maria's utterance and which non-speakers lack. In short, if translationism is right, it behooves us to ask about another's reason for what he believes on the basis of linguistic shapes and sounds (he believes) he perceives. But then, *nothing less than appeal to* other mental states *about* what another perceives can rationalize the belief.

Suppose I'm right about the failure of reliabilism to deliver required rationalizations for beliefs about what is said; and suppose, even further, that such rationalizations require intervening attitudes about the words heard. Question: why can't the translationist rebut: 'OK, but why isn't this just the denial of semantical skepticism, something I've already conceded?' Answer: what I'm assuming is that rationalizations for beliefs parallel rationalizations for actions, that is, both require causally efficacious intervening attitudes. So, a belief that *p* (partly) rationalizes a belief that *q* only if the belief that *p* is (partly) causally responsible for the belief that *q*. Needless to say, this isn't uncontroversial.[10]

There is an important feature of my discussion I have not flagged. Not all that long ago, Chomsky spent too much time defending the psychological/epistemic status of grammatical theories he postulated to account for linguistic comprehension. My argument against translationism circumvents these hairy issues. It's insignificant for the purposes of this debate what the nature of the relationship is between a speaker and whatever metalinguistic information is essential for understanding. If I'm right, then there *must* be a relationship. Whether it amounts to tacit (or implicit) or to explicit knowledge, or whether the relationship is not of knowing but of 'cognizing', or

---

[9] What about adults who lack expressions for meaning and quotation (or anything like them)? Am I claiming that, because they are unable *consciously* to construct a metalinguistic justification for their transitions from what they hear into what is said, it follows that they don't understand their language? No.

[10] Looking back to note 8, if linguistic comprehension is more like thinking than like sensing, it may very well be that the best account for it posits causally relevant intervening internal states. But this is, obviously, something I can no more than mention, and certainly not settle, in this chapter.

whether it is a completely different doxastic relationship is irrelevant. What is essential, if I'm right, is that there must be some such epistemic/psychological/doxastic relationship toward semantic information, which stands between the heard utterance and the acquired belief about what is said, if the latter is not to be an unrationalized psychological state. We can leave open its nature.[11]

Why *semantic* knowledge?

Suppose translationism is wrong. It doesn't follow that knowledge about *meaning* must be invoked to account for transitions from heard utterances to beliefs about what is said. Take Schiffer's story. Why doesn't it suffice to say that what underwrites Harvey's understanding his language is his knowing a *translation* manual from his public language into his language of thought?

A translation manual from L to L′ is a finitely axiomatizable theory that correlates words and structures of L with words and structures of L′ so as to entail theorems that correlate L sentences with their synonyms in L′. Such a theory assigns no semantic values to the expressions of either language and in no sense determines [a meaning theory] for either language [. . .] Harvey works in the following way. His internally represented translation manual determines a function that maps each English sentence onto its Mentalese synonym, and he is so 'programmed' that when he has an auditory perception of an utterance of a, then straight-away there enters his belief box the Mentalese translation of 'the speaker in uttering a said that a' [. . .]. (Schiffer, 1993, p. 244)

It's easy to get snowed by Schiffer's technical jargon and by what appear to be merely heuristic devices, for example appeals to Mentalese, translation manuals, the belief-box, talk about translating public language sentences into Mentalese. We want to know Massimo's reason for his belief that Maria said that it's snowing when he believes she uttered 'sta nevicando'. According to Schiffer, Massimo understands Maria's utterance if he is caused to believe that Maria said that it's snowing (in other words, to token, in his belief-box, a sentence of Mentalese which expresses what 'Maria said that it's snowing' does in English) when he hears her utter 'sta nevicando'. If an epistemic relationship toward some internal psychological state is required for linguistic comprehension, why can't it be knowledge of the correct mapping from the Mentalese counterpart of 'Maria uttered 'sta nevicando'' to a Mentalese counterpart of 'Maria said that it's snowing'?

If this is possible, using L does not require knowing anything about an internally represented *meaning* theory for L. So the inference that knowledge of meaning is required for language comprehension is still not sanctioned. However, elevating Schiffer's suggestion to an object of knowledge doesn't work, seeing that it doesn't show why an adequate account of linguistic comprehension requires reference to

---

[11] So here I depart from Smith (1992).

connections between language mentioned and language used—in other words, *semantic* information.

Schiffer's choices are between a function that maps structural descriptions of Italian expressions into structural descriptions of Mentalese expressions, in which case it is a translation manual (and Davidson's translation argument kicks in); *or* a function that maps structural descriptions of Italian expressions, that is, language mentioned—say, Maria uttered 'sta nevicando'—into language *used*, that is, that Maria said that it's snowing. Everything turns on how we understand the locution a certain sentence is tokened in Massimo's belief-box.[12]

Massimo hears Maria utter a certain sentence and he knows that what she uttered translates into a certain Mentalese sentence. Whatever corresponds in his *lingua mentis* to 'Maria said that' concatenates with this translation, and the entire product goes in his belief-box. That suffices for him to understand Maria's language. What I'm doubting is that we can specify this knowledge in a way that both avoids the standard translation argument and does not itself draw upon semantic information—that is, meaning, truth, or satisfaction conditions.

The standard Fodorian reply that, for Mentalese, questions about understanding cannot arise won't work here. It may be illegitimate to ask: in virtue of knowing what does someone understand one's mentalese? (See Lycan, 1984, p. 237f.) But that is not our question. We are asking: in virtue of what knowledge does one understand his *public* language? The suggestion that it is in virtue of knowing a translation manual from, say, Italian, into Mentalese won't work, if the mapping is from structural descriptions into structural descriptions. But suppose this translation manual maps structural descriptions into, say—well, what? It cannot be propositions or states of affairs. That would be entirely useless to Massimo. What about a sentence used in the language of thought, or a *sentence-in-use* in the language of thought? I want to argue that a 'translation manual' in this sense *determines* a meaning theory, that is, a semantic theory.

Suppose a function F in effect maps a set S of structural descriptions of sentences of L into a set P of sentences (not structural descriptions) of L′ such that F(*s*) (in S) = *p* (in P) iff (if X assertively utters *s* in L, then X says that *p*). Then F determines a semantic theory for L.[13]

T is a truth theory for a language L in a metalanguage L′ if and only if, for every structural description *s* of a sentence of L, T implies a true sentence of L′, such that

s is true in L iff *p*

(where *p* is replaced by a sentence of L′, and each *s* translates whatever replaces *p*). F determines an adequate truth theory T for L in L′ if and only if, for any structural

---

[12] I am disagreeing with Segal (1994, pp. 114–115). Nothing about translationism requires that the function from heard language into the language of thought map structural descriptions onto structural descriptions.

[13] I'll couch my discussion in truth theories, but nothing hangs on this.

descriptions s_1 and s_2 of sentences of L, $F(s_1) = F(s_2)$ only if T implies (for some sentence p of L') (a) and (b):

(a) s_1 is true in L iff p

(b) s_2 is true in L iff p

But $F(s_1) = F(s_2)$ only if there is some p in L' such that both s_1 and s_2 'translate' p. But this establishes that some truth theory for L implies (a) and (b).

If F were merely a translation manual in a traditional sense, this result could not follow. That's what Davidson's translation argument establishes. That 'sta nevicando' translates 'it's snowing' cannot determine that 'sta nevicando' is true if and only if it's snowing. The disquotation principle behind this inference is not innocent. It assumes that 'it's snowing' is true if and only if it's snowing. In the current context that assumption is question-begging, since it's exactly what I'm trying to defend.

None of this establishes, of course, that knowledge of F provides the rationalization which, I claim, (partly) constitutes linguistic comprehension—merely that it *determines* something that could provide this warrant. That I know p, and that p determines something q such that, if I knew that q, my belief that r would be rationalized, do not imply that knowing p alone rationalizes my belief that r. Moreover, even if knowledge of a semantic theory for L suffices for understanding L, it's still open whether knowledge of non-semantic mechanisms might suffice for understanding L as well. So I have not established that semantic knowledge is necessary. And hence, I have not established that semantic theoretical knowledge is necessary, since it is consistent with what I argued that one need not know any more than the meaning theorems that issue from an adequate semantic theory. Massimo's belief that Maria said that it's snowing is justified if he merely believes that 'sta nevicando' means that it's snowing and if Maria assertively uttered 'sta nevicando.' That he need also know that *an object 'a' satisfies 'neve' if and only if 'a' is snow* in order for the transition to be justified requires further argument. These are rather significant loose ends, though I believe they can be tied up. I'll leave that for another occasion.

Conclusion

So where are we? We began with what I still think is the greatest challenge to those of us who find semantics for natural languages not only interesting but valuable, namely how to refute translationism. To this end I argued that translationism leaves unrationalized a speaker's beliefs about what others say. Going externalist about rationalization, as reliabilism recommends, seems misguided, at least to me. Internalist accounts both invoke attitudes about the words we hear and treat those attitudes as causally responsible in effecting beliefs about what is said. Traditional translation manuals mapping structural descriptions into structural descriptions are no help here; and any 'translation' manual that takes words mentioned into words used (or words in use) sneaks in just the semantic information we are trying to redeem. There are obviously many missing steps, and many of the steps taken lack anything like an ironclad defence. But I hope that at least the dialectic is sufficiently precise. Let me end on a different note. The most common reason for resisting the idea that speakers have semantic (theoretical) knowledge is that such knowledge is not 'within the ken of plain folk' (Schiffer, 1987, pp. 255–61), not something to which we have 'conscious access' (Foster, 1975, p. 2), not something we can 'literally credit' speakers with (Dummett, 1974, p. 110). Nothing I'm recommending requires Massimo to have *explicit* representations, or be able consciously to reconstruct pieces of practical reasoning from perceived sounds to extralinguistic belief (Lepore, 1982; Higginbotham, 1983, 1987; George, 1989). Massimo may be unequipped, incapable, or unskilled. But if his beliefs about what's said are rationalized, it makes sense for us to articulate his reasons. My argument, if any good, establishes that some relationship toward metalinguistic states or information about one's language is required for linguistic comprehension. I don't have a clue what the psychological make-up of this relationship must be like. But no one should take a critical stance on these issues without at least having a fairly developed account of concepts against which to evaluate such attributions.

Acknowledgments

I would like to thank Herman Cappelen, Donald Davidson, Ray Elugardo, Lou Goble, John Heil, Kirk Ludwig, Stephen Schiffer, and especially Jerry Fodor and Barry Smith for comments on earlier drafts of this chapter.

References

Baker, G. P., and P. M. S. Hacker (1984) *Language, Sense and Nonsense*. Oxford: Basil Blackwell.
Bruner, J. (1957) On Perceptual Readiness. *Psychological Review* 65: 14–21.
Chomsky, N. (1986) *Knowledge of Language*. New York: Praeger.
Cresswell, M. J. (1978) Semantic Competence. In M. Guenthner Ruetter and F. Guenthner (eds), *Meaning and Translation*, London: Duckworth, pp. 9–27.
Davidson, D. (1967) Truth and Meaning', *Synthese*, 17: 304–323; reprinted in Davidson (2001).
Davidson, D. (1973) Radical Interpretation. *Dialectica* 27: 314–328, reprinted in Davidson (2001).
Davidson, D. (1986) A Coherence Theory of Truth and Knowledge, in *Truth And Interpretation, Perspectives on the Philosophy of Donald Davidson*, Ernest Lepore (ed.), Oxford: Basil Blackwell, 307–319.
Davidson, D. (2001) *Inquiries into Truth and Interpretation*, 2nd edn, Oxford: Oxford University Press.
Davies, M. (1981) *Meaning, Quantification, Necessity*. London: Routledge and Kegan Paul.
Devitt, M. (1981) *Designation*. New York: Columbia University Press.
Dretske, F. (1981) *Knowledge and the Flow of Information*. Cambridge, MA: MIT Press.

Dummett, M. (1974) What Is a Theory of Meaning (I)? In S. Guttenplan (ed.), *Mind and Language*, Oxford: Clarendon Press, pp. 97–138.

Dummett, M. (1978) What do I Know When I Know a Language? in *The Seas of Language*, pp. 94–106.

Dummett, M. (1993) *The Seas of Language*. Oxford: Clarendon Press.

Evans, G. (1981) Semantic Theory and Tacit Knowledge. In S. H. Holtzman and C. M. Leich (eds), *Wittgenstein: To Follow a Rule*, London: Routledge and Kegan Paul, pp. 118–137.

Evans, G., and J. McDowell (eds) (1975) *Truth and Meaning*. Oxford: Duckworth.

Fodor, J. A. (1975) *The Language of Thought*. New York: Crowell.

Fodor, J. A. (1984) *The Modularity of Mind*, Cambridge, MA: MIT Press.

Fodor, J. A. (1987) *Psychosemantics*. Cambridge, MA: MIT Press.

Fodor, J. A. (1989) Why Should the Mind be Modular? In A. George (ed.), *Reflections on Chomsky*, Oxford: Basil Blackwell, pp. 179–202.

Fodor, J. A. (1990) Review of Stephen Schiffers's *Remnants of Meaning*. In J. A. Fodor, *A Theory of Content and Other Essays*, Cambridge, MA: MIT Press, pp. 177–191.

Fodor, J. A. (1991) Reply to Schiffer. In *Meaning and Mind*, B. Loewer and G. Rey (eds), Oxford: Basil Blackwell, pp. 304–310.

Fodor, J. A. and E. Lepore (1992) *Holism: A Shopper's Guide*. Oxford: Basil Blackwell.

Fodor, J. A., J. D. Fodor, and M. Garrett (1975) The Psychological Unreality of Semantic Representations. *Linguistic Inquiry* 6: 515–531.

Fodor, J. A., M. Garrett, E. Walker, and C. Parkes (1980) Against Definitions. *Cognition* 8: 263–367.

Foster, J. (1975) Meaning and Truth Theory. In Evans and McDowell, pp. 1–33.

George, A. (1989) How Not to Become Confused about Linguistics. In A. George (ed.), *Reflections on Chomsky*, Oxford: Basil Blackwell, pp. 90–110.

Goldman, A. (1986) *Epistemology and Cognition*. Cambridge, MA: Harvard University Press.

Hacking, I. (1975) *Why Does Language Matter to Philosophy?* Cambridge: Cambridge University Press.

Harman, G. (1975) Language, Thought, and Communication. In K. Gunderson (ed.), *Minnesota Studies in the Philosophy of Science*, Vol. 7. Minneapolis: University of Minnesota Press pp. 270–298.

Higginbotham, J. (1983) Is Grammar Psychological? In L. Cauman, I. Levi, C. Parsons and R. Schwartz, (eds), *How Many Questions? Essays in Honor of Sidney Morgenbesser*, Cambridge, MA: Hackett Publishing, pp. 170–179.

Higginbotham, J. (1987) The Autonomy of Syntax and Semantics. In J. Garfield (ed.), *Modularity in Knowledge Representation and Natural Language Understanding*, Cambridge, MA: MIT Press, pp. 119–131.

Higginbotham, J. (1988) Is Semantics Necessary? *Proceedings of the Aristotelian Society* 87: 219–241.

Higginbotham, J. (1995) The Place of Natural Language. In P. Leonardi and M. Santambrogio (eds), *On Quine*, Cambridge: Cambridge University Press, pp. 113–139.

Hornstein, N. (1984) *Logic as Grammar*. Cambridge, MA: MIT Press.

Katz, J., and J. Fodor (1963) The Structure of a Semantic Theory. *Language* 39: 170–210.

Larson, R., and P. Ludlow (1993) Interpreted Logical Forms. *Synthese* 95: 305–356.

Larson, R., and G. Segal (1995) *Knowledge of Meaning*. Cambridge, MA: MIT Press.

Lepore, E. (1982) In Defense of Davidson. *Linguistics and Philosophy* 5: 277–294.

Lepore, E. (ed.) (1986) *Truth and Interpretation: Perspectives on the Philosophy of Donald Davidson*. Oxford: Basil Blackwell.

Lewis, D. (1970) General Semantics. *Synthese* 22: 18–67.

Loewer, B., and G. Rey (eds) (1992) *Meaning and Mind: Fodor and his Critics*. Oxford: Basil Blackwell.

Lycan, W.G. (1984) *Logical Form in Natural Language*. Cambridge, MA: MIT Press.

McDowell, J. (1988) In Defense of Modesty. In Barry Taylor (ed.), *Michael Dummett: Contributions to Philosophy* Nijhoff International Philosophy Series, 25. Dordrecht, Holland, and Boston, MA: Nijhoff, pp. 59–80.

Nozick, R. (1981) *Philosophical Explanations*. Cambridge, MA: Harvard University Press.

Partee, B. (1975) Montague Grammar and Transformational Grammar. *Linguistic Inquiry* 6: 203–300.

Partee, B. (1979) Semantics: Mathematics or Psychology? In R. Bäurle and E. Egli (eds), *Semantics from Different Points of View*, Berlin: Springer-Verlag, pp. 1–14.

Quine, W.V.O. (1972) Methodological Reflections on Linguistic Theory. In D. Davidson and G. Harman (eds), *Semantics of Natural Language*, Dordecht: D. Reidel, pp. 442–454.

Richard, M. (1992) Semantic Competence and Disquotational Knowledge. *Philosophical Studies* 65: 37–52.

Schiffer, S. (1982) Intention-Based Semantics. *Notre Dame Journal of Formal Logic* 23/2: 119–156.

Schiffer, S. (1987) *Remnants of Meaning*. Cambridge, MA: MIT Press.

Schiffer, S. (1993) Actual-Language Relations. *Philosophical Perspectives* 7: 231–258.

Schiffer, S. (1994) A Paradox of Meaning. *Nous* 28(3): 279–324.

Segal, G. (1994) Priorities in the Philosophy of Thought. *Proceedings of Aristotelian Society. Supplementary Volume* 68, pp. 107–130.

Smith, B. C. (1992) Understanding Language. *Proceedings of Aristotelian Society* 102: 109–141.

Soames, S. (1989) Semantics and Semantic Competence. In J. Tomberlin (ed.), *Philosophical Perspectives*, Vol. 3: *Philosophy of Mind and Action Theory*, Atrascadero: Ridgeview Publishing Company, pp. 185–207.

Stich, S. (1983) *From Folk Psychology to Cognitive Science*. Cambridge, MA: MIT Press.

Thomason, R. (ed.) (1974) *Formal Philosophy: Selected Papers of Richard Montague*. New Haven, CT: Yale University Press.

Vermazen, B. (1967) Review of Jerrold J. Katz and Paul M. Postal, *An Integrated Theory of Linguistic Description*, and Jerrold J. Katz, *The Philosophy of Language*. *Synthese* 17/3: 350–365.

Wright, Crispin (1986) Theories of Meaning and Speaker's Knowledge. In S. G. Shanker (ed.), *Philosophy in Britain Today*, Albany: State University of NY Press, pp. 267–307.

9

Solipsistic Semantics

Ernest Lepore and Barry Loewer

In a famous passage of the *Meditations*, Descartes writes:

At this moment it does indeed seem to me that it is with eyes awake that I am looking at this paper; that this head which I move is not asleep, that it is deliberately and of set purpose that I extend my hand and perceive it [...] But in thinking over this I remind myself that on many occasions I have been deceived by similar illusions, and in dwelling on this reflection I see so manifestly that there are no certain indications by which we may clearly distinguish wakefulness from sleep that I am lost in astonishment. And my astonishment is such that it is almost capable of persuading me that I now dream. (Descartes, 1967, p. 146.)

In his skeptical arguments, Descartes is claiming not merely that it is possible that all his thoughts about the world are false, but that it is possible for him to have these very thoughts and that they be false. It is possible that he is dreaming that there is a world outside his mind though none exists. Descartes' radical skepticism also involves his view that he can know the contents of his thoughts, even though he knows nothing of the world, or even if there is a world. On the Cartesian picture, the content of a thought is a property intrinsic to the thought and conceptually independent of any individuals outside the mind.[1]

Semantics for a system of mental representations assigns to each mental representation a meaning or content. The assignment can take a number of forms; it can be for example a truth theory, an assignment of Fregean senses, or even an image theory. We will say that semantics is solipsistic (hereafter SS for 'solipsistic semantics') if, in assigning meanings to representations, it does not presuppose the existence of any mental or physical individuals other than the thinker and his thoughts. This characterization needs to be sharpened. There are at least two ways in which an assignment of meanings to representations may presuppose the existence of individuals. One is for the semantics to *interpret* a representation by assigning as its meaning some individual other than the thinker and his representations. For example, semantics that interprets a proper name as directly referring to an actually existing individual implies that the reference of the

---

[1] Some qualification is needed here. Descartes did think that some of our ideas were not conceptually independent of anything outside the mind—for example our idea of God and of infinity.

name exists, or did exist. The other sort of presupposition is a bit more difficult to specify.

Given a language L and an interpretation for L, we can ask the question: in virtue of what does L have that interpretation? For an interpretation of mental representations, the question is: what is it about the representations, their structures and other intrinsic properties, their interactions with each other, with the thinker's physical and social environment, and so forth, that makes them have that interpretation? In virtue of what do they have it? We will call a theory that answers this question for a language L with interpretation I a *theory of meaning* for L. It is important to distinguish the question a theory of meaning is supposed to answer from the question of what events happen to cause a representation to have a particular interpretation. The distinction is parallel to two ways of understanding the question: what makes a good man good? One question is: what causes a man to be good? (or how can we make a man good?). The other question is: what kinds of facts make a good man good? The second question is a conceptual or metaphysical one, about the nature of goodness. It is this kind of question concerning meaning that a theory of meaning attempts to answer.

An example of a theory of meaning that does not presuppose the existence of individuals other than the thinker and his thoughts are certain 'picture' or 'image' theories of meaning.[2] On this view, thoughts are mental images whose representational powers are determined by intrinsic features of the image, for example, its phenomenal color and shape. The image refers to whatever resembles it, but it has the meaning it has entirely in virtue of its intrinsic features. An example of a theory of meaning that does presuppose the existence of individuals external to the thinker is Kripke's causal theory of names. A name means what it names in virtue of bearing a certain causal relation to its bearer. As we pointed out, Kripke's interpretation is also non-solipsistic, since it interprets a name as meaning its bearer. It may be that the correct theory of meaning for some kinds of expressions is solipsistic, whereas the correct theory of meaning for other kinds is non-solipsistic. It is plausible that the mental counterparts of logical constants possess the meanings they do in virtue of their functional roles, and that functional role is solipsistic. SS for L requires that neither the interpretation nor the theory of meaning for L presuppose the existence of individuals external to the thinker.

A theory of meaning and interpretation for L may be solipsistic even though L contains terms that *purport* to refer to individuals external to the thinker. This is just the sort of possibility Descartes envisaged. The central idea of SS is that the determinants of the meanings of one's mental representations are entirely within oneself. A doctrine closely related to SS states that, if two of a person's mental representations have the same or different meanings, then it will be possible for that

---

[2] The view that ideas are images and that images represent by picturing was held by Locke. See his *Essay Concerning Human Understanding*, Books 2 and 3.

person to determine that they have the same or different meanings by introspection alone. We will call this doctrine 'transparency'. Given the accessibility of one's mental representations to consciousness, transparency is entailed by there being SS for mental representations.

In this paper we will examine Cartesian and more contemporary motivations for the view that there must be SS for mental representations. We will discuss some well-known arguments, which show that various kinds of natural language expressions possess a non-solipsistic semantics. An apparent consequence of this state of affairs is that, when sentences containing such expressions are used to specify thoughts, as happens in attributions of propositional attitudes, the thoughts are characterized non-solipsistically. We argue that a semantics for English is so thoroughly non-solipsistic that, even if thoughts have SS, their contents cannot be expressed in English. We then argue that the most plausible theories of meaning for mental representations are also non-solipsistic. Our discussion results in an apparent dilemma. On the one hand, there are the Cartesian intuitions and other motivations for thinking that thoughts possess SS. On the other hand, there are the arguments that seem to show that, given our usual ways of characterizing thought contents, their semantics is non-solipsistic. We conclude with some tentative remarks on how these two views might be reconciled.

Descartes is not alone in his advocacy of SS. Hume remarks that 'to form the idea of an object and to form an idea is the same thing; the reference of the idea to an object being an extraneous denomination, of which in itself it bears no mark or character' (Hume, 1888, p. 20). Hume is saying that the interpretation of his ideas is solipsistic, since one can have an idea of an object even though the object fails to exist. Hume's image theory of meaning is also solipsistic, since it locates the representational powers of an idea in its intrinsic features. Frege also seems to endorse SS. He writes:

Let us just imagine that we have convinced ourselves, contrary to our former opinion, that the name Odysseus, as it occurs in the Odyssey, does designate a man after all. Would this mean that sentences containing the name 'Odysseus' expressed different thoughts? I think not. The thoughts would strictly remain the same; they would only be transposed from the realm of fiction to that of truth. So the object designated by a proper name seems to be quite inessential to the thought-content of a sentence which contains it. (Frege, 1979, p. 191)

Frege is saying that the name 'Odysseus' has the sense it has whether or not Odysseus exists. His account is that to understand the name is to grasp its sense. Although he doesn't say much about what it is to grasp a sense, his view seems to be that the grasping of senses is a matter strictly between the mind and the realm of senses. It requires the existence of no physical or mental individuals (senses are neither mental nor physical) other than the thinker and his thoughts. Frege also upholds 'transparency'. He frequently uses it to establish that two expressions have different senses. Since one can grasp the senses of 'Hesperus' and 'Phosphorus' without realizing that they have the same reference, it follows, according to Frege, that 'Hesperus' and 'Phosphorus' have different senses.

Descartes and Frege were dualists. On their accounts, the mind possesses an intrinsic and unexplained power to represent the world. There are also physicalist versions of SS. Physicalists will attempt to account for semantic facts in terms of physical properties and laws. A physicalist who endorses SS for mental representations thinks that the physical facts that determine the meanings of thoughts involve only intrinsic physical properties of the thinker's body. One can express this view through the claim that the semantic properties of a thinker's mental representations *supervene* on intrinsic physical (for instance neurophysiological) states of her body.[3]

Physicalistic SS can make for strange bedfellows. Two prominent contemporary proponents of it are John Searle and Jerry Fodor, who agree on little else in the philosophy of psychology (Searle, 1980, 1983; Fodor, 1980). Searle expresses SS as follows:

> If I were a brain in a vat I could have exactly the same mental states I have now; it's just that most of them would be false [...] The operation of the brain is causally sufficient for intentionality. It is the operation of the brain and not the impact of the outside world that matters for the content of our internal states. (Searle, 1980, p. 452)

> I think in the relevant sense that meanings are precisely in the head: there is nowhere else for them to be. (Searle, 1980, p. 200)

Searle's view is that the causal operations of the brain are sufficient (and perhaps necessary) to produce thoughts with their contents.

Fodor's version of SS is different from Searle's. Fodor advocates (whereas Searle rejects) the 'computational theory of mind' (hereafter, CTM). According to Fodor, mental processes are analogous to the operations of a computer. For example, when one forms an intention to, say, go to a certain Chinese restaurant for dinner, one's mind (brain) engages in computational processes involving the manipulations of various mental representations. Fodor argues that a great deal of theorizing in cognitive psychology presupposes CTM. He also argues that CTM provides plausible explanations of various features of mental states and processes (for example the opacity of belief). Fodor argues that CTM is committed to the following 'formality condition' (Fodor, 1980, p. 229): if psychological states (processes, and so on) have the same computational characteristics, then they must be the same psychological states (process, and so on). Fodor also maintains that psychological states (or an important subset of them) are characterized in terms of their contents. It follows from this and the formality condition that, if states are computationally the same, then they have the same contents (or rather they are composed of representations with the same contents). In other

---

[3] A property P supervenes on properties Q1, Q2, . . . , if and only if it is metaphysically impossible for two individuals to differ with respect to P without differing with respect to some of the Qs. For a discussion of supervenience, see Kim (1984).

words, thought contents supervene on computational features of thought. Since computational features, whatever they might be, supervene on intrinsic physical features, it follows that thought contents supervene on physical features. In this way Fodor seems committed to the existence of a SS for mental representations.[4]

Fodor's view is more complicated than we have so far indicated. Referring to considerations that we will soon discuss, he observes that our normal attributions of belief violate the formality condition. But he thinks that they come close to satisfying it. He writes that 'taxonomy with respect to content may be compatible with the formality conditions plus or minus a bit' (ibid., p. 250). The suggestion is that, even if our usual scheme of belief attributions assigns contents non-solipsistically, it comes close enough to it, so that, with a little tinkering, we can construct SS for mental representations.

Fodor's reasons for thinking that there exists SS for mental representations are mainly theoretical. He holds that modern cognitive theory requires it. SS also has a powerful intuitive appeal. The intuitions that underlie SS are supported by the Cartesian thought experiment. One can imagine each of his thoughts about the external world being false, even the thought that there is an external world, whereas the thoughts themselves remain the same. So it might seem that thoughts have the contents they have independently of any individuals who are extrinsic to them. Furthermore, although one might be mistaken about the meanings of words in one's public language, it seems absurd to think that one might be mistaken about the meanings of one's own thoughts. Descartes expresses this line of thought beautifully:

Now ideas considered in themselves and not referred to something else, cannot strictly be false; whether I imagine a she-goat or a chimera, it is not less true that I imagine one than the other. [. . .] The chief and commonest error that is to be found in this field consists in my taking ideas within myself to have similarity or conformity to some external object; for if I were to consider them as mere modes of my own consciousness, and did not refer them to anything else, they could give me hardly any occasion of error. (Descartes, 1971, p. 78)

Despite the Fodor's arguments and the Cartesian's intuitions, the very possibility of SS is called into question by recent developments in the philosophy of language and mind associated primarily with work by Kripke, Kaplan, Putnam, and Burge. Taken as a whole, this work apparently shows the inadequacy of the Fregean account of meaning. Because of the close connection between Frege's theory and SS, these developments have a bearing on the possibility of SS for thought.

Saul Kripke's work on names was the spearhead of the attack against the Fregean accounts.[5] According to Frege, a name—for example, 'Aristotle'—expresses a sense that, as it happens, picks out a certain referent (Aristotle). If things had turned out

---

[4] Subsequently Fodor rejected the need for narrow content and adopted a thoroughly referential and externalist semantics and theory of meaning. See Fodor 1990, 1994, 1997, 1998.

[5] Kripke (1980). Kripke does not commit himself to the view that there is a causal relation between the token of a name and its bearer, although others have made this claim.

differently, the same sense might have picked out someone else or nothing at all. Kripke argues that Frege is wrong. A name does not express a sense but, instead, *directly* refers to its bearer. In terms of the apparatus of possible world semantics, a name refers to its bearer at every possible world. A term that expresses a sense—for example, a definite description—may refer to different individuals at different worlds. According to Kripke, the reference of a use of a name is determined by a causal chain that begins with a baptism of the bearer with the name. Someone who hears the name 'Aristotle' used by a competent speaker can himself use it to refer to Aristotle, even if the information that he associates with 'Aristotle' is insufficient to determine Aristotle (e.g. he is a Greek philosopher), or even if the information is uniquely true of someone else (e.g. he is the ancient Greek playwright who produced the trilogy *Oresteia*).

How is Kripke's account of the meaning of names relevant to the semantics for mental representations? It is natural to assume that, when someone is reported as thinking the thought that Aristotle was wise, he is said to have a thought whose content is the same as the content of 'Aristotle is wise'. His mental representation contains a constituent that directly refers to Aristotle. These semantics are non-solipsistic in both ways discussed earlier. The interpretation of 'Aristotle' is Aristotle himself, and so it involves an individual other than the thinker and his thoughts. Also, Kripke's account of what determines the interpretation of a name requires that its bearer exist at the time of baptism. So Kripke's theory of meaning for names is non-solipsistic. We also note that 'transparency' fails for Kripke's semantics. Someone might use the names 'Hesperus' and 'Phosphorus', which, according to Kripke, have the same interpretation, and yet have no way of discovering by introspection that they have the same interpretation. Meaning, according to Kripke, is not entirely in the head.

Kripke's direct reference account of names is superior to Frege's sense theory in a number of ways. However, there are problems with the account that seem to support the view that names must also have a solipsistic interpretation, which is relevant when they are used in contexts ascribing thoughts. If names are directly referential, then, since 'Tully' and 'Cicero' designate the same man, the thought that Tully was an orator and the thought that Cicero was an orator are thoughts with the same interpretation. But it seems possible to believe one without believing the other (Kripke, 1976). If believing that p is to be explained (as in Fodor) as tokening a mental representation that means that p, then it seems that the mental representations corresponding to 'Tully was an orator' and 'Cicero was an orator' must have different interpretations.[6] Another, perhaps even more serious problem is that it certainly seems possible to think that Homer was a Greek even though it turns out that Homer never existed. But on the direct reference account, if Homer never existed, 'Homer was a Greek' would fail to express any proposition.[7] It was precisely consideration of these problems that led Frege to postulate senses as the meanings of names. It is essential to his solution to the

---

[6] Kripke can avoid this objection by abandoning transparency.

[7] Kripke's non-SS requires either that we cannot have such thoughts or that the mental representation corresponding to 'Homer' is not a directly referring expression.

problems that whether a sense determines a reference or whether two senses determine the same reference is irrelevant to the grasping of senses. This suggests that a semantics for names that is adequate for propositional attitude contexts will also interpret them as expressing senses.

David Kaplan has proposed a semantics for indexical sentences that is non-solipsistic (Kaplan, 1989). According to him, when Arabella utters the sentence 'she is a spy', pointing at Barbarella, she asserts a proposition that is essentially about Barbarella. This proposition actually contains Barbarella as a constituent. If it turns out that Arabella is not pointing at anyone, then, on Kaplan's view, her utterance simply fails to express a proposition. It is clear that Kaplan's semantics interprets indexical utterances non-solipsistically.

Suppose that, when Arbella utters 'she is a spy', she expresses a thought that has the same interpretation as her utterance. On Kaplan's account, this thought would contain a constituent that directly refers to Barbarella. According to this account, if Arabella and her neurophysiological twin point respectively at Barbarella and at Twin Barbarella and each utters 'she is a spy', they are thinking different thoughts. So the interpretations of mental representations containing indexicals do not supervene on neurophysiological states.[8]

Even if indexical thoughts have non-solipsistic interpretations, there also seems to be a need to associate solipsistic interpretations with them. From Arabella's point of view, she is thinking the same thought when she utters 'she is a spy', whether or not she is pointing at anyone. If thoughts are individuated in terms of their causal consequences for behaviour, then it makes no difference whether or not Arabella is pointing at Barbarella or at her twin, or is just hallucinating. Kaplan's distinction between the character of an indexical sentence and the proposition expressed by an utterance of the sentence may provide the ingredients for a solipsistic interpretation of indexical thoughts. Character is a function from contexts of utterance to propositions. For example, the character of 'I am in Ann Arbor,' uttered by Arabella at time t, yields the proposition that Arabella is in Ann Arbor at t. This proposition contains Arabella herself as well as Ann Arbor and t as constituents. A semantics that assigns this proposition to indexical thoughts is clearly non-solipsistic. But a semantics that assigns to the thought its character might be compatible with SS. Arabella can have in mind the character of 'she is a spy,' even though she is pointing to no one. Arabella and her twin have the same character in mind.

Character is not sufficiently robust to characterize the semantics of mental representation in all cases.[9] This can be seen in the following situation. Arabella is looking at two

---

[8] One might infer from this non-supervenience that Kaplan holds a non-SS theory of meaning. However, compare note 9. Supervenience fails, and yet SS obtains. Nothing in Kaplan's writings tells whether he is or is not a non-SS theorist of meaning about indexicals. We also note that transparency fails for him. Arabella may be looking in a mirror, pointing unwittingly at herself and thinking what she would express by 'I am tall' and 'she is tall', without knowing that these express the same proposition.

[9] Actually we think something much stronger than this. In Chapter 5 above we argue that, at best, character provides a method for individuating thoughts, not for identifying or ascribing them.

TV screens. One shows Barbarella from the front, the other from the back. Arabella doesn't realize this. She thinks twice 'she is a spy'. It is intuitively clear that her two thoughts are different, even though they share character and express the same proposition. Searle develops an account of indexicals that distinguishes the two thoughts. His view is that the content of Arabella's thought is something like 'the (female) person who is causing *this* visual experience is a spy'.[10] One way of understanding Searle's view is that the proposition expressed by this thought contains the visual experience as a constituent. Searle's semantics is apparently solipsistic, since the interpretation of the thought requires the existence of nothing other than Arabella and her mental contents.

Hilary Putnam first introduced twin arguments (like the one used above) to show that natural kind terms, expressions like 'water,' 'gold,' 'tiger,' and so forth, have non-solipsistic meanings (Putnam, 1975). Arabella and Twin Arabella are neurophysiologically identical. They inhabit, respectively, Earth and Twin Earth, which are identical except for the fact that the stuff called 'water' on Earth is composed of H_2O molecules, whereas the stuff called 'water' on Twin Earth is composed of XYZ molecules. We also suppose that the period of time is one before chemistry was born as a modern science, so that no one on Earth or Twin Earth can distinguish H_2O from XYZ. According to Putnam, Arabella's word 'water' refers to H_2O, whereas Twin Arabella's word 'water' refers to XYZ. A version of the causal theory of reference explains these reference relations. Roughly, Arabella's use of 'water' refers to H_2O because her use is a link in a causal chain that begins with original dubbings of samples of H_2O. Arabella's use refers to anything that belongs to the same natural kind as the original samples. Since XYZ is not the same natural kind as H_2O, Arabella's tokens of 'water' do not refer to XYZ, even though Arabella cannot distinguish the two kinds. If she were miraculously transported to the shores of the Twin Pacific on Twin Earth and said 'water, water, everywhere', she would be wrong.

When Arabella and Twin Arabella think the thoughts that each would express by uttering 'water is wet', they think different thoughts. Arabella's thought is true if and only if H_2O is wet, whereas her twin's thought is true if and only if XYZ is wet. It follows that their thought contents do not supervene on their neurophysiologies. Furthermore, on Putnam's causal account of how 'water' gets its meaning, Arabella could not think her thought unless she was on the receiving end of a causal chain that originates with an event involving H_2O. Putnam's interpretation of, and his theory of, meaning for natural kind terms are squarely nonsolipsistic.[11]

Putnam's positive account is that the meaning of a natural kind term consists of a number of components. One is the reference of the term—in our example, H_2O.

---

[10] Searle (1983, p. 203). Interestingly, on Searle's account Arabella and her twin are having thoughts with different contents, since each refers to her own visual experience. So even though Arabella and Twin Arabella are in type-identical neurophysiological states, they can have solipsistic thoughts with different contents. This shows that we cannot always argue, from the fact that two individuals in type-identical neurophysiological states have thoughts with different contents, to the conclusion that the thoughts are non-solipsistic.

[11] Putnam also abandons transparency. The thoughts expressed by 'water is wet' and by 'H_2O is wet' are the same, but surely one need not recognize them as such.

A second component he calls 'stereotype'. It consists of the information that competent speakers associate with water—for example that it quenches thirst, fills oceans, and so forth. Putnam gives the impression that this component is entirely within the mind. This suggests the possibility that Arabella's mental representation corresponding to 'water is wet' may have two interpretations: a non-solipsistic interpretation, which includes H_2O, and a Fregean solipsistic interpretation, as in 'the stuff that quenches thirst, fills oceans, and so forth, is wet'. We will pursue this idea later.

Tyler Burge pushes a variant of the Twin Earth parables that also shows that the meanings of certain expressions are determined by factors external to the thinker (Burge, 1979). Burge imagines an English speaker who does not know that arthritis is specifically a condition of the joints, although most of her beliefs concerning arthritis are true. She utters 'I have arthritis in my thigh'. According to Burge, her utterance means that she has arthritis (in our sense) in her thigh. His reason for claiming this is that Arabella will defer to members of her linguistic community, should they correct her. When corrected, she will say that her utterance was false. Burge then considers this woman's twin, who speaks Twin English, which is like English, except that in it 'arthritis' refers to inflammations of the thigh as well as of the joints. Her twin's utterance is true. On the assumption that the thought expressed by an utterance has the same interpretation as the utterance, Arabella and her twin think different thoughts even though they are neurophysiologically identical. (This we take to be Putnam's early view.)

It is interesting to compare Burge's and Putnam's arguments. Putnam's argument applies to natural kind terms. Burge's argument can apparently be applied to any expression, even to adjectives, adverbs, and logical connectives. If Putnam is correct, then the interpretation of a natural kind term includes the reference of the term—for example, the substance water. If water never existed, we couldn't think that water is wet any more than we could think that Aristotle is wise if Aristotle never existed. Burge's argument has no such conclusion. The term 'arthritis' might have been introduced by description, and there might never have been cases of the disease.[12] Putnam and Burge have non-solipsistic theories of meaning, but they emphasize different ways in which meaning is determined. Putnam emphasizes the causal connections between the tokening of an expression and a dubbing of a natural kind. Burge emphasizes the role that one's linguistic community has in determining the meanings of one's words. Of course, it may be that both Putnam and Burge are correct and that an adequate theory of meaning for English will include reference both to causal chains and to community practices.

The arguments that we have quickly canvassed purport to show that certain expressions and representations have a non-solipsistic semantics. In each case, the arguments show that these representations have meanings that they could not have

---

[12] This suggests that we could have a language with a non-SS theory of meaning but with SS interpretations.

unless certain individuals other than a thinker and her thoughts existed. How might the view that thought possesses SS be defended against these arguments? One strategy is to fight battles on each front, arguing that Kripke is mistaken about names, Putnam mistaken about natural kind terms, and so forth. This is the strategy pursued by Searle. The other strategy is to admit defeat on the fronts, but then to circle the wagons around the mind and defend the possibility of constructing a characterization of mental content that is solipsistic. This is the strategy pursued by Fodor, and it is the one that we will follow. Still, we can make use of Searle's accounts by taking his views about the correct semantics of English expressions as suggestions for how to construct SS.

Fodor's strategy is to associate with each mental representation a narrow and a wide content (Fodor, NCMH). The arguments of Kripke, Putnam, and others are taken to show that wide content does not supervene on the thinker's body. But narrow content is supposed to supervene on the thinker's body, including on states of his brain and sense organs, and therefore it is a version of physicalistic SS. The problem is to construct an appropriate notion of narrow content.

When he wrote 'Methodological Solipsism Considered as a Research Strategy in Cognitive Psychology', Fodor seemed to think that it would not be all that difficult to construct SS for mental representations that could play an explanatory role in cognitive psychology. After discussing some of the anti-solipsistic considerations we have reviewed, he remarks:

> To summarize: transparent taxonomy is patently incompatible with the formality condition, whereas taxonomy in respect of content may be compatible with the formality condition, plus or minus a bit. That taxonomy in respect of content is compatible with the formality condition, plus or minus a bit, is perhaps *the* basic idea of modern cognitive theory. (Fodor, 1980, p. 240)

Fodor is claiming that, if we stick to opaque as opposed to transparent interpretations of mental representations, we will come close to a semantics that conforms to the formality condition, that is, to SS. But it seems to us that the construction of SS that can be used in cognitive theory is a much more formidable, perhaps an impossible, task. In the remainder of this paper we will consider reasons why this is so.

What are the adequacy conditions that a characterization of narrow content must satisfy? Since it is solipsistic, it will assign the same contents to Arabella's and her twin's thoughts 'water is wet', and it will also assign the same contents to the thoughts of the woman and her twin in Burge's story. It will assign different contents to the thoughts expressed by 'Cicero was bald' and 'Tully was bald' when the thinker does not believe that Cicero = Tully. The characterization of narrow content should serve the needs of cognitive theory. Fodor seems to understand this requirement in such a way that propositional attitudes interpreted narrowly will yield rationalizing explanations of action (when the actions themselves are described narrowly). Folk psychological theory, according to Fodor, contains generalizations like the following: when an individual believes that his obtaining water requires that he raise his hand, and he wants it to be the case that he obtains water, then he will, *ceteris paribus*, raise his hand. This generalization does not apply to twin earthlings, because they do not have beliefs about water. This is due to the fact that belief content is characterized widely in the generalization. Fodor suggests that cognitive theory will contain refinements of such generalizations which apply on Earth and Twin Earth. So a requirement on narrow content ascriptions is that it should be employable in such generalizations. In 'Methodological Solipsism' (1980), Fodor held that the appropriate notion of content satisfies the formality condition and so supervenes on neurophysiological states. In more recent writings, he seems to hold that narrow content supervenes on bodily states, or perhaps on boldily states together with a specification of inputs to the organism's perceptual systems (Fodor, NCMH). If the inputs are characterized in ways that make reference to no individuals external to the organism's body, then the characterization is still a version of SS. One final requirement on narrow content is for there to be a plausible theory of meaning that is solipsistic and that accounts for how representations obtain their contents.

What form will SS take? One currently fashionable answer is provided by 'conceptual role theories' (hearafter, CRT; see for example Loar, 1981; McGinn, 1982; Field, 1977). A CRT for a person's language of thought characterizes the meaning of a mental representation in terms of its causal or inferential role in relating stimuli, behaviour, and the tokening of other mental representations. The characterization is a version of SS only if representations, behaviour, and stimuli are described in ways that make no reference to individuals external to the thinker's body. We have discussed CRTs elsewhere and argued that a characterization of CR is not itself a characterization of content (see Chapter 5 in this volume). For present purposes it is sufficient to point out that a characterization of CR does not yield appropriate complements to put into 'believes that' and other propositional attitude contexts. But we need such expressions of content if we are to construct rationalizing explanations of behavior of the sort Fodor wants to capture in cognitive psychology. It should also be clear that CR does not provide a characterization of content suitable for expressing Cartesian skepticism. Descartes was not claiming that all his thoughts might be false even if they have the same conceptual role as the one they actually have. They could have the same conceptual role and yet be about quite different things. So we have to look elsewhere for a specification of narrow content.

We know of two other proposals for constructing narrow content or SS for linguistic and mental representations. We will call one of these the 'indexicalist strategy'. It involves interpreting thoughts indexically, in a way that is supposed to presuppose no individual external to the thinker. We will call the second approach the 'phenomenological strategy'. It involves finding a collection of expressions that have SS and constructing interpretations for thoughts from these expressions. The approach is called 'phenomenological', since the non-logical vocabulary of these interpretations consists of 'observation' terms that are supposed to describe how things seem or how they appear. In a number of recent papers, Fodor has employed both proposals in order to construct a characterization of narrow content. But we will argue that the prospects for success are poor. We will discuss the phenomenological strategy first.

The phenomenological strategy applied to proper names suggests that a name is interpreted as expressing a sense that the thinker associates with the name. For example, the thought that Aristotle was Greek might be interpreted as having the content that the author of the *Metaphysics* was Greek (or some similar content, which can be expressed without using the proper name 'Aristotle'). The phenomenological strategy might be applied in the following way to natural kind terms. Although Arabella and Twin Arabella refer to different things when they utter 'water is wet', it may be that they associate the same stereotype with 'water'. This suggests that the stereotype has a solipsistic interpretation. The idea is that the thoughts of both twins can be interpreted phenomenologically as having a content like 'the liquid that people drink, fills oceans, and so forth, is wet'.

However, the interpretations that we associated with 'Aristotle was Greek' and 'water is wet' are certainly not completely solipsistic. The description 'the author of the *Metaphysics*' contains another proper name, and so we do not yet have a solipsistic interpretation. And even if the name were replaced by a description, the question would arise whether the predicates that occur in the description can be given solipsistic interpretations. This question also arises when we consider the stereotypes associated with natural kind terms. The suggestion was that the solipsistic interpretation of 'water is wet' is that the liquid that people drink, fills oceans, and so forth, is wet. It is clear that the content of this stereotype is not sufficiently narrow to be solipsistic. The expressions 'oceans' and 'people' have different meanings for Arabella and Twin Arbella. For Arabella, 'people' refers to earthlings, whereas for Twin Arabella it refers to twin earthlings. The same point applies to 'oceans' and 'liquid', and perhaps to other concepts in the stereotype. By imagining suitable differences between Earth and Twin Earth while keeping constant the ways things seem to the twins, it looks as though Twin Earth arguments will succeed in showing that no natural kind term has SS. It might be suggested that those expressions that describe the ways things seem, the truly phenomenological expressions, are immune from the Twin Earth arguments. Are there any such predicates in English? The best candidates are 'observation terms', for example 'red,' 'round', and 'bitter'. If these do not have SS, it is difficult to see how the phenomenological can be made to work.

As we already pointed out, the argument Burge gave to show that the meaning of 'arthritis' depends on features external to the thinker, specifically, community usage, applies to any expression in a natural language. If these arguments are correct, they show that even predicates like 'is red' do not have SS. However, it seems that we can imagine a language that is like English except that the deferential practices on which Burge's arguments rely are absent. For this reason we will give another argument that shows that the semantics of observation terms is non-solipsistic. Suppose that on Twin Earth those things that are red on earth (blood, ripe tomatoes, boiled lobsters—or rather their counterparts on Twin Earth) are green. So, if an earthling visited Twin Earth, she would correctly think that the things that twin earthlings call 'boiled lobsters' are green. However, the twin earthlings are born with colour-inverting lenses, so that, when looking at what they call 'a lobster', they experience the same kind of sensations (or they are in the same brain states) earthlings experience when looking at boiled lobsters. Suppose, as usual, that Arabella and her twin are neurophysiologically type-identical and that each utters 'roses are red'. The things the twin earthlings call 'roses' are actually green.

How should we translate Twin Arabella's word 'red' into English? Her utterances of 'that's a red one' are typically caused by things that are green. When she says 'I am looking for a red dress', she is satisfied when she finds a green one. So we have every reason to suppose that her word 'red' means green. If we translate her word 'red' by our word 'red', the result would be that we would interpret Twin Arabella as being pervasively mistaken about the colours of things. This is certainly intolerable. It is much more plausible to translate her 'red' by our 'green.' This translation interprets that Arabella and Twin Arabella are thinking different thoughts when each says to herself 'that's a red one', even though they are in neurophysiologically identical states and they are experiencing the same qualia.[13]

Fodor might reply that Arabella and her twin are really in identical bodily states, since the twin's colour-inverting glasses count as part of her body. In his most recent discussions of narrow content, Fodor characterizes narrow content so that it super-venes on the states of an organism's brain and transducers (Fodor, NCMH). Plausibly, the colour-inverting lenses are part of the twin's visual system. This reply can be deflected with some more science fiction. We suppose that on Twin Earth a substance in the atmosphere changes light from red to green and vice versa soon after it is reflected. In the revised story, Arabella and her twin are in identical brain and transducer states, although one is thinking 'that's a red one' while the other is thinking 'that's a green one'.[14]

---

[13] We do not mean just that we lack epistemic warrant in translating her word 'red' as our word 'red', but that it would be an error to do so.

[14] Fodor's strategy suggests a fourth semanticist. One might be a SS theorist of meaning but have non-SS interpretations. This depends on several things. First, it is important to see that Fodor is not offering a theory of meaning here. The phenomenological approach is a strategy for ascertaining what thoughts an individual can entertain without looking outside the organism's physical structure. If this strategy succeeded, one might be inclined to infer that there is a SS theory of meaning for mental representations. However, Fodor has not given us one. He is not here telling us *in virtue of* what mental representations have the contents they do. He is telling us only how to discern these contents. Second, Fodor is not a phenomenalist. He does not claim that phenomenological properties are in the head. This is no problem for him. Qua cognitive psychologist, his concern is only that he need not look outside the organism to discern its psychology. However, that these properties exist outside the organism does not commit Fodor to non-SS interpretations. This depends on his ontological view about properties. If properties are universals, abstract entities, then his interpretations are SS. If phenomenological properties are sets of physical objects, then his interpretations presuppose the existence of physical entities outside the organism, and he is therefore committed to non-SS interpretations. A view like this one is developed by Stephen White (1982).

The second approach that the solipsistic semanticist can take is the indexicalist strategy. An indexical sentence has both a content and a character. The character is a function from contexts to contents. So when Arabella and Twin Arabella utter 'she is a spy', their utterances have different contents (one utterance is about Barbarella, the other about Twin Barbarella), but the same character. Character may be solipsistic even if content is not. So, if we can associate an indexical interpretation with each mental representation, we might yet succeed in constructing SS for thought. But we doubt that this can be carried out. Exactly what indexical interpretation can be given to 'water is wet'? Fodor suggests that 'perhaps 'water' means something like 'the local, transparent, potable, dolphin-torn, gong tormented [. . .] stuff one sails on' (Fodor, BD. The reference to its being local provides the indexicality. Arabella and her twin may mean the same by 'water' and yet refer to different substances, because they inhabit different contexts. But the obvious problem with this suggestion is that this paraphrase contains expressions that are interpreted non-solipsistically.

Searle suggests an interpretation for 'water' that may seem to avoid this problem. He says that ' "water" is defined indexically as whatever is identical in structure with the stuff causing *this* visual experience' (Searle, 1983, p. 203). But this suggestion faces a couple of difficulties. First, it is not clear that 'stuff' and 'causing', or even 'visual experience,' have a non-solipsistic semantics. Second, the expression does not uniquely refer. There are many things and events causing *this visual experience*. One of the causes is the pattern of neuron firing in the optic nerves. We could exclude this by adding 'the external liquid stuff causing [. . .]', but now we are faced with the problem that 'external liquid' does not have SS.

There is another way to pursue the indexicalist strategy.[15] Think of Arabella's entire environment, including its history, as a context. Her sentence 'water is wet' can be interpreted as expressing a character that maps that context onto the content that H_2O is wet. Twin Arabella's sentence expresses the same character, which maps her different context onto the content that XYZ is wet. It is clear that character construed in this way supervenes on neurophysiology. But it cannot serve as a specification of narrow content. At most, the account provides sufficient conditions for when two individuals' thoughts have the same meaning, and then, only for when the two are neurophysiologically identical. If two people are not neurophysiologically identical, then the account says nothing concerning whether structurally similar representations possess the same or different characters. Nor does this characterization of meaning as character yield appropriate specifications of content that can follow 'believes that'. As Fodor, who suggests this proposal, says: 'first it is one thing to have a criterion for the intentional identity of thoughts; it is quite another to be able to say what the intention of a thought is' (Fodor, BD). It is the latter that we need if we want to employ narrow content in the rationalizing explanations of cognitive psychology.

---

[15]   A view like the one following is developed in White, 1982.

The preceding considerations show that it is not easy to construct SS. The usual, nonsolipsistic interpretation of thoughts misses being solipsistic by a great deal more than 'plus or minus a bit'. There is no solipsistic part of English out of which solipsistic interpretations of thought can be constructed. But this doesn't show that SS is impossible. It shows only that our language is so thoroughly non-solipsistic that it doesn't contain the resources to construct a non-solipsistic semantics. Perhaps mental representations have SS that cannot be expressed in English. However, we will argue that the most plausible theories of meaning for mental representations rule out SS. Since Fodor himself advocates a version of this theory of meaning, our argument will be, to a certain degree, an ad hominem one.[16]

The account that Fodor favours is a development of Fred Dretske's views. According to Dretske, the content of a mental representation is determined by its informational origins (Dretske, 1981, Ch. 10–11). To consider the simplest kind of case, suppose that a mental structure can either be in state Y or in state N and that, under certain 'normal conditions', it is in Y when the organism is looking at a fly but in N when the organism is not looking at a fly. Then the mental structure carries the information that there is a fly in view when conditions are normal. Fodor endorses an account of this kind, although he substitutes certain epistemically ideal conditions for normal conditions (Fodor, P).

Although there are substantial difficulties with Dretske's and Fodor's proposals, it is the most promising one.[17] Here we only want to show that, if we take this account seriously, then either it results in non-solipsistic semantics or it leads to SS that interprets one's mental states as being about one's own nervous system. First, observe that, on a plausible understanding of 'normal' and 'ideal', Fodor's theory of meaning is already non-solipsistic. Normal and ideal conditions are such in virtue of the evolutionary history of an organism. If the organism had evolved in different circumstances, different conditions might count as normal or ideal. Let's ignore this problem for a moment. Suppose that a frog's mental state carries the information that a fly is in view. Clearly it carries this information only because of its interactions with its environment. Change the environment, and the frog's neural state will carry different information, for example, that a moving BB is in view. The obvious reply to this point is to say that the frog's neural state was designed by evolutionary pressures to carry information about the presence of flies, not of BBs. But the proponent of SS cannot avail himself of such considerations, because the self-assigned task is to construct meaning from ingredients that are entirely within the head. Neurophysiologically identical organisms might have evolved in different environments, and so their neural states might carry different information. If we are to use a Dretskean theory of meaning to support a

---

[16] We are not claiming that Fodor ever intended Dretske's informational theory of meaning to be a *solipsistic* theory of meaning of these representations. However, this is the only kind of theory of meaning that Fodor has presented for mental representations.

[17] Informational accounts of content are discussed by Barry Loewer, in Loewer (1987).

solipsistic interpretation of the frog's mental representations, then we will have to find something for these representations to carry information about that is immune from the kind of environmental tinkering used in the Twin Earth stories. But it would seem that all that can remain constant under these environmental changes is the working of the frog's nervous system itself. There will be some co-variation between states Y and N and certain patterns of irradiation on the frog's eyes. So we can solipsistically interpret Y as having the content that a certain pattern of occular irradiation is occurring. If we adopt a Dretskean theory of meaning, then the only interpretations consistent with SS interpret an organism's thoughts as being about its own nervous system.

A proponent of SS holds that, although thoughts and ideas purport to refer to the external world, their meanings are entirely a product of mental activity. Originally he may have thought that the semantics for all the expressions of our language is solipsistic. But the considerations advanced by Kripke, Putnam, and Kaplan show that vast portions of a natural language have interpretations that are incompatible with SS. This showed that our usual ways of individuating the contents of mental representations are also non-solipsistic. We considered a response that granted the points made by Kripke and others, but we attempted to construct solipsistic interpretations out of the fragment of language that remained solipsistic. We argued that this strategy is unlikely to succeed because almost all natural language expressions, even observation terms, have a non-solipsistic semantics. The prospects for constructing SS for thought seem even bleaker when we reflect on the difficulty of providing a plausible account about what makes it the case that a thought has the content it has, which is compatible with solipsism. If an informational answer to this question, one of the sort suggested by Dretske and Fodor, is correct, then a language with SS is a language that is solipsistic in another way. In it one can refer only to oneself and to one's mental states. Solipsism with respect to sense results in solipsism with respect to reference.[18]

If the view that thought possesses SS is not plausible, what are the consequences for cognitive theory and for our Cartesian intuitions? With respect to cognitive theory, there seem to be two alternatives. The first, recommended by Stich (1983), is to abandon the use of propositional attitude explanations in cognitive theory. The second, recommended by Burge (1986), is to accept the fact that propositional attitudes do not supervene on bodily states and to argue that this in no way counts against their explanatory ability. We cannot enter into this debate here. But we do want to make a few remarks concerning the Cartesian intuitions that seem to provide such strong support for there being SS for thought.

---

[18] This is not necessarily a criticism. Someone might simply swallow the idea that all our thoughts are about our own mental states in order to keep meaning in the head. But Fodor cannot do so, if he is to provide rationalizing explanations of the sort discussed above. No matter how narrowly we describe Arabella's and Twin Arabella's behavior, we will not be able to construct a cognitive psychology that issues in rationalizations of their behavior if we limit these twins to having thoughts only about their own brain states.

The Cartesian observes that it is possible for all his thoughts about the external physical world to have the contents they have even if there is no external world. If this is correct, then thought–world interactions cannot be an essential determinant of the contents of thoughts. It follows that thought has SS. The first point to make about this argument is that it shows, at most, that it is *epistemically* possible for the Cartesian to have the thoughts he actually has even if there is no external world. The argument is similar to other Cartesian arguments that show that it is epistemically possible for a mental event or state to exist even though no physical events or states exist. But epistemic possibility is not the same as metaphysical or conceptual possibility, and only these would establish the need for SS. However, we must admit that our arguments do not establish that SS is not possible. At most, we have shown that plausible theories of meaning are non-solipsistic.

There is a related intuition, which may lead the Cartesian into thinking that thought must have SS. It is the observation that one can know the contents of one's thoughts without engaging in any empirical investigation. When considering another person, we might be persuaded that his thoughts have the contents they have in virtue of interactions with his environment, his linguistic community, and so forth. But I can know the contents of my own thoughts without evidence and, in particular, without knowing much about these matters. Since I can know the contents of my thoughts without investigating the world outside myself, it is tempting to conclude that these contents must be determined by events entirely within myself. We have here a real and, we think, quite deep tension. The tension between non-solipsistic theories of meaning and the Cartesian intuitions would be relieved somewhat if it could be shown that there is no genuine conflict between the claim that we know our own thought contents without evidence and the claim that those thought contents are determined by matters external to us, about which we may have no knowledge.

We would like to show that, even though a mental representation R has the content it has in virtue of matters external to the thinker, the latter might still know R's content without evidence. We tentatively offer the following account. Let us suppose that A grasps the concept water. We cannot give necessary and sufficient conditions for grasping this concept, but something like the following story seems plausible. To grasp the concept water is to have a mental representation that plays a certain kind of functional role in one's thought and is related in an appropriate way to H_2O.[19] Suppose that A thinks the thought that water is wet. What is required of A in order that he know what his thought means? If it is required that he knows that water is H_2O, then it must be admitted that he does not know what his thought means. But this requirement is not plausible and does not follow from the fact that A grasps the concept. Now suppose that A grasps the concept 'is true' and has a way of referring to his thoughts—say, by quoting them. A's thought 'water is wet' is true iff water is

---

[19] This kind of account of concepts is developed in Colin McGinn (1981).

wet will then express his knowledge of the content of his thought 'water is wet'. And A will know that this thought is true simply in virtue of grasping the concepts of *truth*, *quotation*, *water*, and *is wet*. Suppose that A had lived on Twin Earth. In that case, his concept would be different, since it would refer to XYZ. However, he could know the content of his thought 'water is wet', since he knows the truth conditions. He would express this knowledge by saying ' "water is wet" is true iff water is wet'.

It might be objected that in our example A does not know that 'water is wet' is true if and only if water is wet, but only that the representation ' "water is wet' is true iff water is wet' is true.[20] The source of the objection is this. Suppose that B knows how quotation works in English, knows the disquotational effect of 'is true', and is able to recognize grammatical sentences of English but knows nothing else about English. B will be in a position to recognize English sentences of the form ' "water is wet' is true iff water is wet' as true. But this is not the same as knowing the truth conditions of 'water is wet'.[21] Someone could recognize this sentence as true and yet not know what it means. The objection is that A is in the same position with regard to his thoughts as B is with respect to English. Our reply is that there is an important difference between the two cases. We assumed that A had the ability to think that water is wet on the basis of there being a mental representation that plays a certain conceptual role in his thinking and on the basis of its playing this role in a particular environment. The sentence 'water is wet' does not play the appropriate functional role in B's thought, which it would if B understood 'water is wet'. But A's thought 'water is wet' does play the appropriate role; it must, if it is to be the thought that water is wet. If this is correct, then we can see how the meaning of a person's thoughts can depend on matters external to him, and yet how he can be in a position to know the meaning of a thought without appealing to these external matters.

Acknowledgments

Earlier versions of this paper were read at the universities of Alberta, Calgary, Oklahoma, Michigan, Central Michigan, Regina, and the Florence Center for the History and Philosophy of Science. We would like to thank all the members of these various philosophy departments for their helpful suggestions and criticisms. We would also like to thank Paul Boghossian, Donald Davidson, Umberto Eco, Peter Klein, Hilary Putnam, Steven Shiffer, and Bas van Fraassen for discussion on earlier drafts of this paper.

---

[20] This objection was made to us in conversation by Paul Boghossian, who remains unpersuaded by our reply.
[21] Much is made of this distinction in Chapter 1 of the present volume.

References

Burge, T. (1979) Individualism and the Mental. *Midwest Studies* 4: 73–121.

Burge, T. (1986) Individualism and Psychology'. *Philosophical Review.* 3–45.

Descartes, R. (1967) *The Philosophical Works of Descartes*, Vol. 1, tr. and ed. Elizabeth S. Haldane and G. R. T. Ross. Cambridge: Cambridge University Press.

Descartes, R. (1971) *Descartes' Philosophical Writings*, tr. and ed. E. Anscombe and P. Geach. New York: Prentice Hall.

Dretske, F. (1981) *Knowledge and the Flow of Information*. Cambridge, MA: MIT Press.

Field, H. (1977) Logic, Meaning, and Conceptual Role. *Journal of Philosophy* 7: 379–409.

Fodor, J. (1980) Methodological Solipsism Considered as a Research Strategy in Cognitive Psychology. In J. Fodor, *Representations*, Cambridge, MA: MIT Press pp. 225–256.

Fodor, J. (1987) *Psychosemantics: The Problem of Meaning in the Philosophy of Mind*, Cambridge, MA: MIT Press.

Fodor, J. (1990) *A Theory of Content and Other Essays*, Cambridge, MA: MIT Press.

Fodor, J. (1994) *The Elm and the Expert: Mentalese and Its Semantics*, The 1993 Jean Nicode Lectures, Cambridge, MA: MIT Press.

Fodor, J. (1998) *Where Cognitive Science Went Wrong*, The 1996 John Locke Lectures, Oxford: Oxford University Press.

Fodor, J. (n.d.) Banish Discontent [= BD]. Unpublished manuscript.

Fodor, J. (n.d.) Narrow Content and Meaning Holism [= NCMH]. Unpublished manuscript.

Frege, G. (1979) *Posthumous Writings*. eds. H. Hermes, F. Kambartel, and F. Kaulbach, Basil Blackwell, Oxford.

Hume, D. (1978) *A Treatise on Human Nature* [1739], ed. L. Selby-Bigge. Oxford: Oxford University Press.

Kaplan, D. (1989) Demonstratives. In J. Almog, J. Perry, and H. Wettstein (eds), *Themes from Kaplan*. New York: Oxford University Press, pp. 481–504.

Kim, J. (1984) Concepts of Supervenience. *Philosophy and Phenomenological Research* 45: 153–176.

Kripke, S. (1979) A Puzzle about Belief. In Margalit (ed.), *Meaning and Use*. Dordrecht: Reidel, pp. 239–283.

Kripke, S. A. (1980) *Naming and Necessity*. Cambridge, MA: Harvard University Press.

Loar, B. (1981) *Mind and Meaning*. Cambridge: Cambridge University Press.

Locke, J. (1979) *Essay Concerning Human Understanding* [1690], ed P. Nidditch. Oxford: Oxford University Press.

Loewer, B. (1987) From Information to Intentionality. *Synthese* 70: 287–317.

McGinn, C. (1982) The Structure of Content. In A. Woodfield (ed.) *Thought and Content*. Oxford: Oxford University Press, pp. 207–258.

Putnam, H. (1975) The Meaning of 'Meaning'. In H. Putnam, *Mind, Language, and Reality: Philosophical Papers*. Cambridge: Cambridge University Press, Vol. 2, pp. 131–171.

Searle, J. (1980) Minds, Brains, and Programs. *The Behavioral and Brain Sciences* 3: 417–424.

Searle, J. (1983) *Intentionality*. Cambridge: Cambridge University Press.

Stich, S. (1983) *From Folk Psychology to Cognitive Science*. Cambridge, MA: MIT Press.

White, S. (1982) Partial Character and the Language of Thought. *Pacific Philosophical Quarterly* 63: 37–65.

10

A Putnam's Progress

Ernest Lepore and Barry Loewer

> In the medieval ages, before the arrival of scientific thinking as we know it today, well, people could believe anything, anything could be true. [...] But the wonderful thing that happened was that then in the development of science in the western world certain things came slowly to be known and understood. I mean, you know, obviously all ideas in sciences are constantly being revised. [...] That's the whole point. But we do at least know that the universe has some shape and order. And that trees do not turn into people and goddesses and there are very good reasons why they don't and you can't believe absolutely anything.
>
> Louis Malle, *My Dinner with Andre*

Hilary Putnam tells us that metaphysical realism (henceforth, MR) has been the dominant view in philosophy in every historical period—at least until Kant (Putnam, 1978, p. 1). One can find it in Plato, Aristotle, Descartes, and Locke. In this century, Russell and early Wittgenstein (on the usual interpretation) were explicit advocates of it, and currently Michael Devitt (1981, 1984), Hartry Field (1972, 1982), Clark Glymour (1982), and David Lewis (1986) confess to it. Putnam himself says he held it implicitly, if not explicitly (1983, p. vii). However, Putnam has announced that he has discovered MR is incoherent (1978, p. 124), and suggests that we should replace MR with another kind of realism, internal realism (IR), a realism with a human face. Putnam argues that MR is committed to a picture of the relation between thought (or language) and reality that is fundamentally implausible; it leads to an unacceptable view of the nature of truth and to pernicious dichotomies. On the other hand, IR is claimed to maintain the scientific aspects of realism while rejecting its metaphysical aspects. So Putnam has given us a characterization of MR, has produced a bill of indictment against it, and has sketched an alternative view, IR. We think it fair to say that Putnam has not succeeded in convincing many metaphysical realists of the errors of their ways. The usual reaction to his arguments is puzzlement. It has seemed to many that Putnam has betrayed his previous realist contributions—he was even pronounced a renegade by a former student (Devitt 1984)—and that he has adopted in its stead a soft (some would say a quasi-mystical) view. Our aim here is to relieve some of this puzzlement by constructing a 'reading' of Putnam's progress from metaphysical to internal realism.

Scientific and metaphysical realism

Bas van Frassen characterizes scientific realism (SR) in this way: 'science aims to give us, in its theories, a literally true story of what the world is like; and acceptance of a scientific theory involves the belief that it is true' (1980, p. 8). Putnam would agree with this characterization and would add that we sometimes have good reasons to believe that scientific theories, particularly those in the mature sciences, are true or approximately true and that the entities they posit exist. The truth of a theory is independent of our beliefs concerning it. Our present good reasons for believing a theory to be true do not guarantee that it is true or approximately true. Any theory we presently hold we may come to reject for good reasons. Furthermore, successive theories can often be viewed as better approximation to the truth. Thus Bohr's 1934 theory of the electron is closer to the truth than his 1912 theory. Such comparisons require interpreting terms within successive theories as referring to the same entities—in this case, to electrons. One of Putnam's most important contributions to the defense of SR was to show how a causal account of reference can be used to support claims concerning intertheoretic reference. Bohr's 1912 term 'electron' and his 1934 term 'electron' were appropriately causally related to the same entity. Of course, this claim is made on the basis of present theory (and what we know of the history of the development of Bohr's views) and is itself defeasible. Putnam considers SR to be an overarching *empirical* hypothesis which best explains scientific practice and success (1978, 123).

We will not attempt to assess SR or Putnam's arguments for it.[1] Rather, we want to show how naturally it seems to fit into a more general philosophical view concerning the relationship between language and the world, namely metaphysical realism. Putnam's official characterization is as follows (1981, p. 49):

(1) The world consists of a *fixed* totality of mind-independent objects and properties.
(2) Truth involves some sort of correspondence relation between words or thought-signs and external things and sets of things.
(3) There is exactly one true and complete description of the way the world is (though we may never have a language capable of expressing it or may never know it).

MR appears to lend philosophical support to SR by providing an account of truth and reference that applies to all theories. It allows for the possibility that terms in different theories refer to the same theory-independent entities. By characterizing truth non-epistemically (in terms of correspondence), it allows for the possibility that even our best theories might yet turn out to be false. And it seems to provide a way of

---

[1] For a detailed assessment of SR and Putnam's arguments for it, see Devitt, 1984.

making sense of the idea that successive theories converge on the one true description of reality.

But, according to Putnam, there are important differences between MR and SR. He says that MR 'is supposed to apply to all theories at once [...] and the world is supposed to be independent of any particular representation of it' (1978, p. 125). On the other hand, scientific realist claims—for example that Bohr's two uses of 'electron' referred to the same phenomena—are made within a particular theory, present-day quantum theory. Putnam's alleged discovery is that MR, instead of supporting SR, actually undermines it. As we will see, the heart of his argument is that the reference and correspondence relations invoked by MR cannot be placed with a scientific account of the world.

The arguments against MR

In this section we discuss some of Putnam's arguments against MR. He observes that

the most important consequence of metaphysical realism is that *truth* is supposed to be *radically non-epistemic*—we might be 'brains in a vat' and so the theory that is 'ideal' from the point of view of operational utility, inner beauty and elegance, 'plausibility', simplicity, 'conservativism', etc., *might be false*. 'Verified' (in any operational sense) does not imply 'true', on the metaphysical realist picture, even in the ideal limit. (Putnam, 1978, p. 125)

Putnam's claim is that MR's (1), (2), and (3) entail:

  (4) Truth is radically non-epistemic.
  (5) It is possible that we are all brains in a vat even though we believe we are not.
  (6) Even an epistemically ideal theory may be false.

(5) and (6) are Putnam's glosses on (4), and he appears to take (5) to imply (6). The argument proceeds:

I assume THE WORLD has (or can be) broken into infinitely many pieces. I also assume T_1 *says* there are infinitely many things (so in *this* respect T_1 *is* 'objectively right' about THE WORLD). Now T_1 is *consistent* (by hypothesis) and has (only) infinite models. So by the completeness theorem (in its model theoretic form), T_1 has a model of every infinite cardinality. Pick a model M of the same cardinality as THE WORLD. Map the individuals of M one-to-one into the pieces of THE WORLD, and use the mapping to define relations of M directly in THE WORLD. The result is a satisfaction relation SAT—a 'correspondence' between the terms of L and sets of pieces of THE WORLD—such that the theory T_1 comes out *true*—true of THE WORLD—provided we just interpret 'true' as TRUE (SAT). So what becomes of the claim that even the ideal theory T_1 might *really* be false? (ibid., pp. 125–126)

The structure of the argument is this:

   (i)   MR implies (6).
   (ii) But an epistemically ideal theory has an interpretation SAT which makes it true.
   (iii) So, the epistemically ideal theory is true.
   (iv) So, MR is false.

There is a number of places where one might take issue with the argument. It is not obvious that the concept of an ideal theory is clear.[2] And it may be objected that, although 'even an ideal theory is false' is consistent with MR, it is not obvious that MR *entails* that it is possible for an ideal theory to be false.[3] But the place where most will balk at Putnam's argument is the move from (ii) to (iii). If someone told us that a theory, for example, the phlogiston theory, is true because it is consistent and so has a model in THE WORLD, we would think him crazy. We would point out the theory's defects: it makes false predictions. The case of an ideal theory is different, since such a theory has no epistemic defects, for example it fails to predict an observation (Putnam, 1978, pp. 126–7; 1981, pp. 45–8, 55; 1983, pp. 8–15). But the defender of MR may still object that SAT is not the correct interpretation of the ideal theory; the correct interpretation SAT★ is the one determined by the reference and correspondence relations between the terms of T_1 and THE WORLD. Interpreted by SAT★, T_1 may turn out to be false. Now Putnam asks this question: What makes SAT★ *the correct interpretation*? This is the key question. The force of the argument is to claim that MR must answer it if Putnam is to defend his view. We will consider answers he might give shortly. First, we want to consider a different response to the argument, which seems to avoid the need for answering the question. Suppose that a version of MR argues as follows: I cannot say what makes a particular reference relation the correct one, but I can show that, no matter what it is, it allows for the possibility that even an ideal theory may be false. Here is how (Putnam, 1978, p. 125):

   (a) It is possible that I am a brain in a vat experiencing the stimuli I experience even though I believe I am not a brain in a vat.
   (b) The ideal theory that is epistemically ideal for a brain in a vat that experiences the same stimuli that I experience is identical to the epistemically ideal theory for me.
   (c) Among the statements included in the ideal theory is that I am not a brain in a vat.
   (d) But in that case the ideal theory constructed by the brain in the vat must be false.

---

[2] Putnam has himself on several occasions remarked that the notion of an (epistemically) ideal theory is far from clear (1983, pp. 12, 161).
[3] MR does not mention 'ideal' theory. So it is hard to see how it could imply 'it is possible that an ideal theory is false'. But, since MR is supposed to provide a non-epistemic characterization of truth, it does seem that the MR would embrace (6). See Glymour, 1982, p. 176.

This response to Putnam's argument, whatever its other features, depends on the assumption that a brain in a vat can entertain the thought that it is a brain in a vat. But Putnam has argued that, given the causal theory of reference, it is not possible for a brain in a vat to even entertain the thought that it is a brain in a vat (Putnam, 1981, Ch. 1). Putnam's argument for this conclusion is usually taken to be a new (or not so new) argument against skepticism, but critics are quick to observe that as such it is not very convincing, since the argument depends on a premise that the skeptic is unlikely to allow: the causal theory of reference (Brueckner, 1986, Conee, 1987, van Inwagen 1988). But the defendant of MR whom we imagine making this response to Putnam's original argument accepts the causal theory of reference. If we see the brain-in-a-vat argument as a reply to the response of that version of MR, then Putnam's puzzling article seems to make sense.

In 'A Problem about Reference', Putnam constructs an argument that is related to, but slightly different from, the argument based on Godel's completeness theorem. Putnam calls this the permutation argument: 'I shall argue that even if we have constraints of whatever nature that determine the truth value of every sentence in every possible world, still the reference of individual terms remains indeterminate' (Putnam, 1981, p. 33).

Suppose that there is a language L and an assignment of truth values to each of the sentences of L, relative to a possible world w, These assignments might be determined by whether or not the sentences satisfy certain constraints, for example, that of being part of an ideal theory at that world. Putnam shows that there are distinct interpretations of the predicates and constants of the language that agree on the assignment of truth values at possible worlds. He takes this to show that no view about how content is determined which works simply by assigning truth conditions to whole sentences is capable of explaining the reference relation. We seem to have the following argument:

(i′) To defend MR one must specify the reference relation.
(ii′) Specifying the reference relation by listing a collection of sentences even if we take those sentences to have truth conditions—functions from possible worlds to truth values—is not sufficient for specifying a unique reference relation.
(iii′) So, MR cannot be defended.

A natural response to both of Putnam's 'model-theoretic' arguments is to say that the interpretations of one's terms and thought-signs are fixed by their causal connections to items in the world. In fact, a number of commentators have expressed puzzlement as to why Putnam, who is one of the developers of the causal theory of reference, does not simply recognize that a causal theory of reference will succeed in specifying the 'intended' interpretations (Blackburn, 1984, p. 301; Brueckner, 1984, p. 137; Devitt,

1984, pp. 86–87; Glymour, 1982, p. 177; Harman, 1982, pp. 569, 573; Leeds, 1978, p. 113; Lewis, 1986):[4]

The natural modern reply to the rhetorical question [What fixes reference?] is that all of the intended interpretations should be replaced by talk of causally determined reference relations; roughly, our physical and social circumstances, and sometimes perhaps our beliefs as well, determine together a set of links, connecting words and objects, and thus delimiting the admissible interpretations of our theories. (Glymour, 1982, p. 177)

So Glymour's view is that the admissible interpretation of our language, the interpretation under which an ideal theory might turn out to be false, is determined not by the sentences of the theory but by causal relations between bits of the world and terms of the theory.

Putnam's response to the suggestion that a causal theory of reference will succeed in answering the key question has exasperated his critics. He says that the causal theory of reference is just more theory and observes that the argument applies to it as well; in particular, there are interpretations that assign 'cause' some relation cause★, which makes all the sentences in the causal theory of reference (and in the rest of the theory) come out true. Putnam thinks that MR is begging the question by appealing to the causal theory of reference (or to any other account, for that matter), since the appeal works so as to single out a unique reference relation only if we assume that 'cause' refers to cause—and not, say, to cause★. Glymour, Devitt, and others think that at this point Putnam has moved from name-calling to game-playing. Their claim is *not* that adding the causal theory of reference to their theory fixes reference, but rather that causation itself fixes reference. The causal account of reference also is supposed to apply to the terms appearing in it—so that the reference of 'cause' is causally determined to be cause instead of cause★.

---

[4] Carsten Hansen writes:

That Putnam in 'Models and Reality' does not discuss his previous work in 'The Meaning of 'Meaning'' is quite puzzling. For in the latter, after having argued that 'meanings aren't in the head', he goes on to give a realistically acceptable account of the extensions in a nonsolipsistic, social setting. [...] It needs to be explained why Putnam no longer thinks this explanation is available to a realist.' (Hansen, 1987, p. 91)

There are two reasons why the causal theory of reference, which Putnam *continues* to hold, does not provide a sufficient response to the challenge. The first is that the causal theory as developed by Putnam provides at most constraints on what counts as an acceptable interpretation and it itself uses intentional notions and so presupposes that the problem of specifying an interpretation has already been solved. For my use of 'water' to refer to water on Putnam's theory requires that I have certain perceptions and intentions—for example, that I intend to refer to the same substance that whoever I acquired the term from used it to refer to. The second response concerns a particular feature of reference. On some versions of the causal theory of reference, certain terms in my language—names and natural kind terms—can refer only if there is a causal chain connecting my use of the term with its reference. This causal chain may go via other speakers. (It is not clear that Kripke holds the view that there must be such a causal chain.) This requirement seems much too weak to single out a unique interpretation, or even to remove from consideration the interpretation SAT, which makes the ideal theory true.

Devitt accuses *Putnam* of begging the question:

> The question begging is most striking in Putnam's latest response to the idea of a causal theory of reference. He claims that 'if reference is only determined by operational and theoretical constraints then the reference of terms in that theory of reference will themselves be indeterminate'. [...] Maybe so but if reference is determined causally, as the theory says it is, then the reference of those terms will be determinate. He is not entitled to assume the theory is false in order to show it false. (Devitt, 1984, p. 190)

Devitt has missed the force of Putnam's argument that there are alternative reference relations that result in identical assignments of truth values in every possible world. If there is a causal characterization of reference, there also are alternative characterizations (in terms of cause★, cause★★, and so on) that agree on the truth values of statements at all possible worlds. The question Putnam asks of MR is: why say that the account of reference in terms of cause rather than in terms of cause★ is the correct one? On MR, both accounts cannot be correct, since presumably the fact that x causes y is a different (physical) fact than the fact that x causes★ y (even though the sentences are true in the same worlds), and the fact that MR says that x refers to y is a particular fact in the world. So the challenge to MR is to give some reason for holding one identification of reference rather than another. The problem that confronts the naturalistically minded defender of MR (as those who put forward causal accounts claim to be) is that, from the point of view of scientific explanation, the different identifications of reference are equally acceptable. That is, if the causal theory assigns truth conditions in such a way as to meet whatever standards are placed on the theory, so will the causal★ theory. Putnam is not, as Devitt (1984, p. 189) says, like a small child delighted in the discovery that there is no end to questions. Instead, he has asked a question—why identify reference with cause rather than cause★?—which the naturalistic MR cannot answer.[5] Further, this question is not unanswerable because the exponent of MR lacks some added piece of information. It is in principle unanswerable as long as the exponent of MR keeps his empiricist (naturalist) credentials. If one were to give up naturalism, one could reply to Putnam that it simply is a brute metaphysical fact that reference is identified with causation and not with causation★. Here name-calling—it is a magical, mysterious theory of reference—would be warranted (Putnam, 1981, p. 3). For this would be to suppose that an *intrinsic* property of an object can determine its relation to a particular object external to it.

We can now summarize our discussion of the model-theoretic arguments as follows: none proves that MR is false. Rather, they are vivid ways of asking an upholder of MR what, on his view, determines the reference of representations. That Putnam also

---

[5] When I ask why reference is fixed by causation and not by causation★, the only answer a physicalist can give me is: 'because that is the nature of reference'. To say that Nature itself singles out objects and put them into correspondence with words is a claim that has no meaning that I can make out at all (Putnam, DIR).

sees his arguments in this light is shown by the following remark: 'The only paper in this book which makes use of technical logic, "Models and Reality" is not an attempt to solve the problem [how correspondence is fixed] but rather a verification that the problem really exists.' (1983, p. ix).

Putnam and Kripkenstein

MR's picture of the relation between language and the world is this:

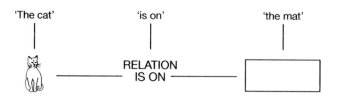

The vertical lines indicate reference relations between terms and parts of the world. The important feature is that the holding of these relations are themselves facts—facts that must obtain for 'the cat is on the mat' itself to succeed in stating a fact. Putnam's question to MR, then, is: what fixes these reference relations? Or, in what do these facts of reference consist? For an upholder of MR who is also a naturalist, it is important that in answering this question he makes use only of naturalistic properties and facts. To complete Putnam's argument against MR, one would have to show that such a proponent of MR cannot provide an answer to this question that is consistent with his naturalistic scruples. Although Putnam does not complete the argument, it seems to us that the situation is exactly this: reference and meaning are *not* naturalistically reducible. Kripke argues for this claim in his discussion of Wittgenstein on rule-following (1981).

The question Kripke asks is similar to the one we have found in Putnam. What fact makes it the case that my use of 'plus' refers to plus (or means that plus in Kripke)? Kripke observes (as does Putnam) that citing either images in my mind when I use 'plus' or my intentions to use 'plus' in a particular way cannot provide a satisfactory answer, since images and intentions are themselves open to interpretation. The heart of Kripke's objection is that, if it were a fact that my use of 'plus' means plus, then that fact would have to possess a certain normativity. It would have to make it the case that, if I answer 'five' to the question 'what is 68 plus 57?', I would be *wrong*. Kripke argues that no "natural" fact possesses this normativity. For example, the fact that I have dispositions to answer various questions involving "plus" is not the appropriate kind of fact, since the dispositions do not themselves distinguish right answers from wrong ones. One might think that a causal theory of reference might be of use here.

A kind of causal theory suggested by Fodor (1987) and Stalnaker (1984) is this: a predicate P (in the language of thought of a person X) has as its extension, say, cats if and only if, under optimal conditions, X's tokenings of P are caused by cats [already in iff]. There are many difficulties with this kind of account (see Loewer, 1987). Here we will discuss two. One is that, to make this account even slightly plausible, the specification of 'optimal conditions' will make ample use of intentional notions, including specifications of X's other beliefs. Obviously this is inadmissible in a naturalistic reduction of reference. Second, a Goodmanesque problem arises: if, under optimal conditions, tokenings of P are caused by all and only by cats, they will also be caused by all and only by cats*, where y is a cat* if and only if, if conditions are optimal, y is a cat and, if conditions are not optimal, y is a dog.[6] Although neither Kripke nor Putnam has produced a proof that a reduction of semantic facts to natural facts is impossible, given the preceding it seems an extremely poor bet. If we are right in our reading of Putnam, he has noticed the same problem that Kripke has for MR. And the prospects for the MR proponent answering the question seem extremely dim.

Must the MR proponent solve the Kripke problem to defend his views? He might make a familiar move toward the view that cat* is a peculiar property and really is not, other things being equal, an eligible referent for a term. This amounts to saying that there is a special class of 'natural' properties that are better candidates than the gruesome properties to be the referents of our terms.[7] In the context of a naturalistic version of MR, this is an odd view. What makes something a gruesome or natural property? That the natural properties just happen to be the ones we refer to might be taken to mean we simply *read* the structure of our language into nature.

A comparison with moral statements is useful. A naturalistically minded MR considers moral statements as non-fact-stating, unless he can show how they are made true by natural facts. If he cannot do so, he will consider them as performing some other function—say, of expressing preferences or attitudes (for example Ayer, 1936; Mackie, 1977). Is a similar move open to the MR defender? The following argument suggests that it is not (Wright, 1984; Boghossian, RF). Suppose that S is true if and only if p. If 'S is true' expresses an attitude, not a fact, then it seems to follow that 'p' does not state a fact either. For the MR defender, this is to throw the baby out with the bathwater.

---

[6] A slightly different theory is found in Dretske 1981: If the original tokens (those that occurred during what Dretske calls the 'learning period') of the predicate P carried the information 'cats are present', then subsequent tokens (those that occur subsequent to the learning period) refer to cats even if they are not caused by cats. The trouble with this is that the appropriate notions of information and the learning period will likely include intentional concepts. See Loewer, 1987.

[7] David Lewis (1986) has taken up the suggestion that there are certain classes of things 'out there' that are intrinsically distinguished, and he suggests that it is a 'natural constraint' on reference—that is, a constraint that is built into nature—that as many of our terms as possible should refer to these classes. Lewis's natural constraint is not brought into existence by our interests. It has to be thought of as something that operates together with those interests to fix reference (Putnam, DIR).

So, although an MR proponent might be led to relegate ethics to a second-class status, he cannot do the same for semantics without undermining his own position.

We would like to consider briefly two other responses open to the adept of MR. One is just to embrace semantical relations and facts as irreducible (or primitive). Putnam reserves his most creative name-calling for this position, accusing it of being 'mediaeval' and 'magical'. It amounts, for Putnam, to saying that the mind somehow reaches out (perhaps by shooting noetic rays!) and touches objects. Beyond the name-calling, there is the irony that a position originally intended as a philosophical basis for scientific realism would end up appealing to non-natural facts.[8] The second response is to try to mitigate the mysteriousness of non-natural facts by claiming that they *supervene* on natural facts. But, as Simon Blackburn has argued in the case of the alleged supervenience of moral properties on natural properties, without a reduction of the former to the latter, 'supervenience becomes, for the realist, an opaque, isolated logical fact, for which no explanation can be preferred' (1971, p. 119) The problem is that, if moral statements are realistically interpreted as stating facts, supervenience is itself a mystery, unless there is a reduction of moral properties to natural ones. One can respond by abandoning moral realism while keeping supervenience as a constraint on the attribution of moral predicates. As we saw, this response is not available to the adept of MR. So, as long as he cannot produce a reduction of semantical properties to natural ones, the supervenience claim does little to dispel the mystery.

Internal realism

Putnam tells us that it was Dummett who awakened him from his metaphysical slumber, and it will be useful to describe briefly certain features of Dummett's account of meaning and truth as preparatory to discussing IR. Dummett, like Putnam, rejects the view that truth is radically non-epistemic. According to Dummett,

A statement is true only if we have established it as true; or only if we either have done so or shall do so at some future time; or only if we have some procedure which, were we to carry it out, would establish it as true; or, at least, only if there exists something of the sort that we normally take as a basis for the assertion of a statement of that class, such that, if we knew it, we should treat it as a ground for the truth of the statement. (Dummett, 1981, p. 443)

This idea is best illustrated in Dummett's philosophy of mathematics (1978, xxiv). Associated with a mathematical statement are conditions for proving it. *Understanding*

---

[8] In some places Putnam makes the stronger charge that this view, of taking reference as primitive, is unintelligible. In 'Models and Reality', Putnam says that this move is unintelligible only if you resist making certain kinds of moves, which some do not resist (Putnam, 1983, p. 14). For example, Chisholm, following Brentano, contends that the mind has a faculty of referring to external objects which he calls 'intentionality', The MR Putnam is criticizing, though, is naturalistically minded, and would find the postulation of an unexplained mental faculty epistemologically unhelpful and almost certainly bad science as well (ibid., p. 5).

the sentence is a practical ability to manifest a particular sort of behaviour, namely, that behaviour which brings us into the position in which, if the condition that conclusively justifies the assertion of S obtains, we *recognize* it as so doing. In the case of a mathematical statement, this requires that we recognize what would count as a proof of the statement. The recognizable conditions of a sentence's conclusively justified assertion are its verification conditions. Dummett denies that a sentence has truth conditions other than those associated with it recognitionally by the speaker's under-standing. So, a sentence has only *verificationist* truth conditions. A mathematical sen-tence, thus, is 'true' if and only if it has a proof (Dummett, 1973, pp. 467–478; 1975, pp. 115–123; 1976, pp. 70–111).

All understanding for Dummett is verificationist. Although Dummett has not developed a worked-out account for non-mathematical sentences, he thinks the situation for these sentences is similar. To understand, for example, 'there is a chair in the room' is to know what would count as a (conclusive) justification of it: perhaps seeing a chair in the room under good light conditions, not being intoxicated, and so on.[9] On this view, there may be sentences such that neither they nor their negations are true. (To justify the negation of S is [conclusively] to refute S.) We may know what counts as a (conclusive) justification (and refuation) of a sentence S, but none might exist. Finally, on this account, reality is mind-dependent (at least) in this sense: whether the sentence 'there is a chair in the room' is true depends on its justifiability conditions. Truth and rationality (justifiability) are conceptually bound—no truth without conclusive justifiability. For example, our finite abilities limit what we can justify.

Dummett's view contains numerous subtleties. Here we want to ignore most of them and pursue a line of thought that will be useful for our comparison with Putnam.[10] It is notorious that Dummett wavers between considering 'justification' to mean 'conclusively justified' and 'sufficiently well justified' (1973, pp. 146, 467, 514, 586; 1978, p. xxxviii). The problem is that, on first reading, it may be that only mathematical and sense-perception statements, if even they, are ever justified. If truth is identified with conclusive justification, it would follow that only mathematical and sense-perception statements are capable of having truth value. Also, since for Dummett the justification conditions of a sentence are its meaning and since understanding a sentence is just knowing its meaning, it follows that anyone who understands a sentence knows its justification conditions, that is, the canonical conditions such that, if they were satisfied, the sentence would be true. An obvious difficulty with this view is that it makes it extremely unlikely that two different people, especially different people who live at different times or in different societies, speak languages that are

---

[9] Dummett is well known for his subscription to the dictum 'a theory of meaning *is* a theory of understanding' (1975, p. 99; 1976, pp. 69–70).

[10] See Dummett, 1978, p. xxxviii, where he articulates various worries about the plausibility of his antirealism.

translatable. For example, it is extremely unlikely that any sentence spoken by Tycho Brahe has the same justification conditions as our sentence 'the sun revolves around the earth'. The result is incommensurability. We would like to say that Tycho Brahe was wrong to believe that the sun revolves around the earth, but Dummett's theory, together with the fact that our procedures for verifying or confirming sentences change, seems to preclude us from doing so.

With this capsule summary of Dummett under our belts, we can describe IR by contrast. Putnam's own proposal, IR, is very sketchily presented. The heart of his view seems to be a rejection of the metaphysical realist picture. It is replaced with a view that truth is justification. He writes:

My own view [...] is that truth is to be identified with justification in the sense of *idealized* justification, as opposed to justification on the present evidence. (Putnam, 1983, p. xvii)

Consider the sentence 'there is a chair in my office right now'. Under sufficiently good epistemic conditions any normal person could verify this, where sufficiently good epistemic conditions might, for example, consist in one's having good vision, being in my office now with the light on, not having taken a hallucinogenic agent, etc. There is no single general rule or universal method (*contra* Dummett) for knowing what conditions are better or worse for justifying an arbitrary empirical judgement. (Putnam, 1983, p. xvi)

Putnam is not trying to give a formal definition of truth, but an informal elucidation of the notion. The two central ideas of his idealization theory of truth are:

(a) Truth, though independent of justification here and now, is not independent of all justification. To claim a statement is true is to claim that it could be justified under ideal conditions.

(b) Truth is expected to be stable or convergent; if both a statement and its negation could be 'justified', even if conditions were as ideal, as one could hope to make them, there would be no sense in thinking of the statement as having a truth value. (Putnam, 1981, p. 56).

His view seems to be this: our practices of forming judgments, testing them, arguing about them, and so on are sufficient for associating with statements what would count as reasons for believing the statements. For some statements, we have a pretty clear idea of what it would take to justify them and the conditions under which they would be justified; for example, 'there is a chair in my office right now'. Putnam offers no account of knowledge, but presumably he has in mind that there is some account of better and worse reasons for my believing that there is a chair in my room right now. Somehow, community practices result in associating justification conditions with statements. These are conditions such that, if we believed they obtained, we would be justified in stating that X believes that *p*. Since truth is idealized justification, it is not the case that 'the sun revolves around the earth' changed truth value and/or meaning between the time of Tycho Brahe and now. It is rather the case that views have changed about whether that statement is (ideally) justified. This is an improvement on Dummett's views, but it is committed to the position that it is by definition impossible for our ideally justified beliefs to be false. We saw before that, despite the metaphysical realist's protests, Putnam argues that the former cannot maintain his intuition about the logical possibility of the ideal theory being false. IR does not rule out the possibility that we might be very well justified (ideally) in believing someone else to be a brain in a vat. But IR seems to rule out the possibility of any individual thinking of herself as a brain in a vat and also being (ideally) justified in this thought.

It is important to clear up certain misconceptions about IR, which are perhaps due to Putnam's own rhetoric. In denying the metaphysical realist (1) (above, p. 47), Putnam is denying that the world consists of a fixed totality of mind-independent objects. This smells of 'idealism', and Putnam's favourite metaphor, that the mind and the world together make up the mind and the world, encourages the charge that IR has idealist tendencies. One way of expressing these tendencies would be to say that, under IR, ordinary commonsense counterfactuals fail to be true. But it is no consequence of IR that counterfactuals like 'if we had not constructed the theory of electrons, then there would be no electrons' are true. In fact, on Putnam's—but perhaps not on Dummett's—account, the counterfactual 'even if we had not constructed the theory of electrons there would be electrons' is justified. So we have reason to think it is true. In general, whenever S is justified, so is 'even if I had not thought of S it would be justified'. (Of course, this counterfactual is not justified for every S.) This may not satisfy a proponent of MR, since he wants to say not merely that one would be justified in asserting the counterfactual, but that it is a *fact* that, even if I had never thought of electrons, they would have existed. But the proponent of IR *is* justified in saying this as well. So, whatever the alleged mind-dependence of the world consists in for the IR, it cannot be expressed by him counterfactually, since he and the MR agree on (most) counterfactuals they assert.[11] The difference, rather, is in what they think the truth of these counterfactuals consists in. For the proponent of MR, it is correspondence to facts. For the proponent of IR it is in being ideally justified.

As early as 1978, Putnam made some rather strong statements about the relationship between realism and equivalent descriptions (1978, pp. 50–51). He stated that we cannot ignore the existence of pairs of equivalent descriptions. Scientific realism is not committed, he contends, to there being one true theory (and only one). He says that 'assuming there is a "fact of the matter" as to "which is true" if either whenever we have two intuitively "different" theories is naieve'. Putnam's intuition is well accommodated by his IR. Since idealized justification is relative to theory, then, for example, relative to field theory 'Fields are real' may be ideally justified, though relative to particle theory it may not be ideally justified. Therefore, it is consistent with IR that there should be more than one 'true' theory or description of the world (*pace* the

---

[11] Peter van Inwagen (1988) for reasons unspecified seems to think otherwise.

metaphysical realist's proposition (3)). It is not surprising, therefore, that Putnam should be charged with relativism (Pears, 1982). Putnam is a relativist inasmuch as whether a statement is true is relative to a theory. But this is a very limited relativism,. Putnam claims that ordinary statements about ordinary objects—for example 'there is a chair in the room'—are, if true, true relative to all acceptable theories (Putnam, NIR). Relativism applies only to statements on the periphery, for example the one about fields above, or statements in set theory (See Putnam 1978, pp. 38–41; 1981, pp. 117–119). Whether Putnam's 'relativism' is this confined or not, Putnam argues that IR does not possess what he takes to be the two most pernicious features of relativism: (1) relativism cannot distinguish between '*p* is true' and 'I (we) think that *p* is true'; and (2) relativism leads to rampant incommensurability. We cannot go here into why Putnam's views escape these charges, except to remark that (a) there is a distinction on IR between ideally justifying 'I think that *p*' and ideally justifying '*p*' (Putnam, 1981, p. 124), and (b) as we have mentioned a number of times, Putnam thinks that translation claims between theories can themselves be ideally justified and that we frequently have good reason to believe these claims are satisfied (ibid., 116–117).

The obvious question that one is itching to ask Putnam is this: doesn't a problem analogous to MR's problem of explaining how the reference relation is fixed arise for IR? What makes it the case that a particular sentence has the justification conditions it has, as opposed to other justification conditions? Similar questions could be asked concerning the reference and meanings of terms. The MR defender may smugly suppose that the proponent of IR will have as much difficulty as he had in specifying the naturalistic facts that make these semantical relations hold. This way of asking the question reveals an MR bias. The metaphysical realist is asking what *fact* makes it the case that 'there is a chair in the room' has whatever ideal justification conditions it has. But the internal realist has rejected the account by which facts make assertions true. This being so, he need not satisfy the metaphysical realist's demand in order to defend his theory. (In contrast, the metaphysical realist did need to satisfy the internal realist's request, since the metaphysical realist held that there is a fact about that which terms refer to.) What he needs to do instead is to show that the sentence 'it is part of the ideal justification conditions of "there is a chair in the room" that' is itself justified—and, similarly, for sentences like '"Cat" refers to cats' and 'Bohr's term "electron" in 1912 has the same reference as his term in 1934'. Someone may ask what fact makes it the case that these semantical sentences have the justification conditions they have. The IR answer is that there is no such fact. Rather, certain assertions about their justification conditions are justified. As Putnam says (quoting Dummett), 'facts are soft all the way down' (Putnam, 1978, p. 128).

The internal realist can reply to the permutation argument as follows: he says that '"Cat" refers to cats, not to cats★' is justified. He can support this by pointing out that 'cat' is appropriately causally related to cats but not to cats★. Neither objection made against the metaphysical realist applies here, since the internal realist is *not searching for a naturalistic fact* with which to identify reference.

Conclusion

The IR's response to the Kripkenstein problem may appear to be a sleight of hand. Simply say that sentences have justification conditions, not realist truth conditions, and—voila!—the problem of intentionality vanishes. But does it? Isn't there, for example, still a problem of accounting for the causal efficacy of intentional (and other mental) states, if these are not naturalistically reducible? These are questions we address elsewhere (see Chapter 11 of this volume). Here we rest, content to have uncovered what we think to be a provocative 'interpretation' of Putnam's text.

Acknowledgments

We would like to thank Bruce Aune, Michael Hand, Peter van Inwagen, Tim Maudlin, Eva Piccardi, Michael Root, and especially Hilary Putnam for comments on earlier drafts of this paper.

References

Ayer, A. J. (1936) *Language, Truth and Logic*. London Dover Publications.

Blackburn, S. (1973) Moral Realism. In J. Casey (ed.), *Morality and Moral Reasoning*,. London: Metheun, pp. 101–124.

Blackburn, S. (1984) *Spreading the Word*. Oxford: Oxford University Press.

Boghossian, P. (n.d.) A Survey of Recent Work on Rule Following [= RF]. Unpublished manuscript.

Brueckner, A. (1984) Putnam's Model-Theoretic Argument against Metaphysical Realism. *Analysis* 44: 134–140.

Brueckner, A. (1986) Brains in a Vat. *Journal of Philosophy* 83: 148–167.

Conee, E. (1987) Review of Hilary Putnam, *Reason, Truth and History. Nous* 21: 81–95.

Devitt, M. (1981) *Designation*. New York: Columbia University Press.

Devitt, M. (1984) *Realism and Truth*. Princeton: Princeton University Press.

Dretske, F. (1981) *Knowledge and the Flow of Information*. Cambridge, MA: MIT Press.

Dummett, M. (1973) *Frege: Philosophy of Language*. London: Duckworth, and Cambridge MA: Harvard University Press.

Dummett, M. (1974) What Is a Theory of Meaning (I)? In S. Guttenplan (ed.), *Mind and Language*, Oxford: Clarendon Press, pp. 97–138.

Dummett, M. (1976) What is a Theory of Meaning? (II). In G. Evans and J. McDowell (eds), *Truth and Meaning*, Oxford: Oxford University Press, pp. 67–138.

Dummett, M. (1978) *Truth and Other Enigmas*. London: Duckworth, and Cambridge MA: Harvard University Press.

Dummett, M. (1981) *The Interpretation of Frege's Philosophy*. London: Duckworth, and Cambridge, MA: Harvard University Press.

Dummett, M. (1983) Language and Truth. In R. Harris (ed), *Approaches to Language*, Oxford: Pergamon, pp. 95–125.

Field, H. (1972) Tarski's Theory of Truth. *Journal of Philosophy* 69: 347–375.

Field, H. (1982) Realism and Relativism. *Journal of Philosophy* 79: 553–567.

Fodor, J. (1987) *Psychosemantics*. Cambridge, MA: MIT Press.

Glymour, C. (1982) Conceptual Scheming or Confessions of a Metaphysicalist Realist. *Synthese* 51: 169–180.

Hansen, C. (1987) Putnam's Indeterminacy Argument: The Skolemization of Absolutely Everything. *Philosophical Studies* 51: 77–99.

Harman, G. (1982) Metaphysical Realism and Moral Relativism. *Journal of Philosophy* 79: 568–574.

Kripke, S. (1981) *Wittgenstein on Rules and Private Language*. Cambridge, MA: Harvard University Press.

Leeds, S. (1978) Theories of Reference and Truth. *Erkenntnis* 13: 111–129.

Lewis, D. (1986) Putnam's Paradox. Unpublished manuscript.

Loewer, B. (1987) From Information to Intentionality. *Synthese* 70: 287–317.

Mackie, J. L. (1977) *Ethics. Inventing Right and Wrong*. New York Viking Press.

Pears, D. (1982) Review of Putnam's *Reason, Truth and History, London Review of Books*, 4/10 (June): 11–12.

Putnam, H. (1975) The Meaning of 'Meaning.' In K. Gunderson (ed.), *Minnesota Studies in the Philosophy of Science*, Vol. 7, Minneapolis: University of Minnesota Press, pp. 131–193.

Putnam, H. (1978) *Meaning and the Moral Sciences*. London: Routledge and Kegan Paul.

Putnam, H. (1981) *Reason, Truth and History*. Cambridge: Cambridge University Press.

Putnam, H. (1983) *Realism and Reason. Philosophical Papers*, Vol. 3. Cambridge: Cambridge University Press.

Putnam, H. (n.d.) A Defense of Internal Realism [= DIR]. Unpublished manuscript.

Putnam, H. (n.d.) A Note on Internal Realism [= NIR]. Unpublished manuscript.

Stalnaker, R. (1984) *Inquiry*. Cambridge, MA: MIT Press.

van Fraassen, B. (1980) *The Scientific Image*. Oxford: Oxford University Press.

van Inwagen, P. (1988) On Always Being Wrong Midwest Studies in Philosophy 12: 95–111.

Wright, C. (1984) Kripke's Account of the Argument against Private Language. *Journal of Philosophy* 81: 759–777.

11

Mind Matters

Ernest Lepore and Barry Loewer

> Who knows what I want to do? [. . .] Isn't it all a question of brain chemistry, signals
> going back and forth, electrical energy in the cortex? [. . .] Some minor little activity
> takes place somewhere in this unimportant place in one of the brain hemispheres
> and suddenly I want to go to Montana or I don't want to go to Montana. Maybe it's
> just an accidental flash in the medulla and suddenly there I am in Montana and I find
> out I really didn't want to go there in the first place [. . .] It's all this activity in the
> brain and you don't know what's you as a person and what's the brain and what's
> some neuron that just happens to fire or just happens to misfire [. . .]
>
> Don DeLillo, *White Noise*

Consider the following, admittedly imprecise claims:

(1) The mental and the physical are distinct.
(2) The mental and the physical causally interact.
(3) The physical is causally closed.

Much can be said in favor of each of these. In support of (1), we can point to the failure
of attempts to reduce the phenomenal and the intentional to the physical, and to
arguments from Descartes to Donald Davidson which purport to show that such
reductions are, in principle, impossible. (2) is supported by our everyday experience
and by various theories of perception and action. (3) means that every physical event or
fact has, in its causal history, only physical events and facts. Both (3) and its cousin:

(3′) All causation is reducible to, or grounded in, physical causation

—where 'grounded' means, roughly, that causal relations supervene on non-causal
physical facts and laws—have seemed to many philosophers to be supported by the
development of the sciences.

The trouble is that it seems that (1), (2), and (3) are incompatible. To be a bit more
definite, consider their application to events. (1) then says that no mental event is a
physical event; (2), that some mental events cause physical events and vice versa; and
(3), that all the causes of physical events are physical events. The inconsistency is
obvious. If mental events are distinct from physical events and sometimes cause

them, then obviously the physical is not causally closed. The dilemma posed by the plausibility of each of these claims and by their apparent incompatibility is, of course, the mind–body problem.[1]

Our primary concern here is how Davidson's[2] account of the relation between the mental and the physical, which he calls 'anomalous monism' (AM), attempts to resolve the dilemma. AM consists of the following three theses:

(4) There are no strict psychophysical or psychological laws, and in fact all strict laws are expressed in a purely physical vocabulary (the anomalousness of the mental).
(5) Mental events causally interact with physical events.
(6) Event c causes event e only if there is a strict causal law which subsumes c and e (entails that c causes e) (the nomological character of causality).

(4) is a version of (1). It is commonly held that a property expressed by M is reducible to a property expressed by P (where M and P are not analytically connected) only if there is an exceptionless bridge law that links them.[3] So it follows from (4) that (intentional) mental and physical properties are distinct.[4] (6) says that c causes e only if there are singular descriptions D of c and D′ of e and a strict causal law L such that L and 'D occurred' entail 'D caused D′' ('Causal Relations', p. 158 (see above, n. 2)). (6) and the second part of (4) entail that physical events have only physical causes and that all event causation is physically grounded.[5]

The notion of a law being *strict* figures prominently both in Davidson's affirmation of the distinctness of the mental and the physical and in his account of causation. Davidson's notion of a strict law is best explained by contrast with non-strict laws. A non-strict law is a generalization that contains a *ceteris paribus* qualifier, which specifies that the law holds under 'normal or ideal conditions', where the relevant notions of normal or ideal are specified by the theoretical context of the law. The generalizations one finds in the special sciences are mostly of this kind. In contrast, a strict law is one that contains no *ceteris paribus* qualifiers; it is exceptionless, not just *de facto* but as a matter of law. A non-strict law may be improved upon by explicitly including some of its *ceteris paribus* conditions in its antecedents. Davidson's view is that psychophysical laws of the form 'whenever a person is in physical state P, then he is in intentional state

---

[1] Similar characterizations of the mind–body problem can be found in Mackie, 1979 and Skillen, 1984.
[2] This view is given in three places in Davidson, 1980: at the beginning and end of 'Mental Events', pp. 208, 223, and in 'Psychology as Philosophy,' p. 231. Unless noted, all references to chapter titles (accompanied only by page references) in the text are to this book.
[3] Davidson's argument against psychophysical laws is restricted to laws whose psychological predicates express propositional attitudes.
[4] We shall typically speak of features, aspects, and properties of events. For present purposes, however, unless we indicate otherwise, what we say can be recast in terms of events satisfying descriptions or predicates.
[5] Davidson never provides an example of a strict causal law. And there are some philosophers who think that his account of causation is much too stringent, because there may be too few strict causal laws. (The best candidates for such laws are basic laws of quantum mechanics.) It is not our aim here to defend Davidson's metaphysical account of causation.

M' are *essentially* non-strict. That is, no matter how many conditions are added to the antecedent, short of trivializing the generalization, it will not be strict.[6]

Given the parallel between (4)–(6) and (1)–(3), it may seem that the former are also incompatible. But they are not. Davidson shows that they all can be true if (and only if) mental events are identical to physical events ('Mental Events', p. 215). Let us say that an event *e* is a physical event just in case *e* satisfies a basic physical predicate (that is, a physical predicate appearing in a strict law). Since only physical predicates (or predicates expressing properties reducible to basic physical properties) appear in strict laws, it follows that every event that enters into causal relations satisfies a basic physical predicate. So, those mental events which enter into causal relations are also physical events.

AM is committed only to a partial endorsement of (1). The mental and the physical are distinct insofar as they are not linked by strict law—mental properties are not reducible to physical properties; but they are not distinct insofar as mental events are physical events. This being so, one might wonder whether AM also only partially endorses claims (2) and (3). In fact, Davidson's views have been criticized precisely on the point of (2). Ernest Sosa (1984) writes: 'I conclude that [. . .] anomalous monism is [not] really compatible with the full content of our deep and firm conviction that the mind and body each acts causally on the other' (p. 278). Ted Honderich goes even further, charging that AM is really a form of epiphenomenalism: 'I went on [. . .] to claim that [AM] was epiphenomenalist; it did not make the mental as mental an ineliminable part of the explanation of actions' (1984, p. 88).

If Honderich means that Davidson's views are committed to epiphenomenalism with respect to mental events, he is clearly mistaken, since, according to AM, mental events do cause other events. They *are* physical events and so can, like any event, have consequences. It is rather that, on AM, as he puts it, the mental *as* mental (some writers use the equivalent expression 'qua mental', or the expression '*in virtue of being* mental') is causally irrelevant. In defence of Davidson, one might reply that, although it is correct to say that it is not *c* as mental that causes *e*, this has nothing to do with any epiphenomenalism on the part of the mental, but simply reflects the fact that it is not events *as* mental or *as* physical or *as* anything else that cause other events. Causation is a relation between events, not between events *as Fs*. It seems to Davidson's critics, however, to make sense to distinguish some features of an event as causally relevant and others as causally irrelevant. It is this distinction that underlies the locution that it is *c* as F (not as F') that causes *e* (to be G).

Ernest Sosa and Fred Dretske[7] illustrate their understanding of the distinction in the following passages, respectively:

---

[6] For an explication and defense of Davidson's arguments for the impossibility of strict psychophysical laws, see Kim, 1986 and McLaughlin, 1986.

[7] Dretzske, 1989. For similar characterizations and examples of causal relevance, see Honderich, 1982, p. 61; Searle, 1983, pp. 155-157; Anscombe, 1975, p. 178; and Achenstein, 977, p. 368.

A gun goes off, a shot is fired and it kills someone. The loud noise is the shot. Thus if the victim is killed by the shot it is the loud noise that kills the victim. [...] In a certain sense the victim is killed by the loud noise. Not by the loud noise as a loud noise but only by the loud noise as a shot, or the like. [...] The loudness of the shot has no causal relevance to the death of the victim. Had the gun been equipped with a silencer the shot would have killed the victim just the same. (Sosa, 1984, pp. 277–278).

Meaningful sounds, if they occur at the right pitch and amplitude, can shatter glass, but the fact that these sounds have a meaning is irrelevant to their having this effect. The glass would shatter if the sounds meant something completely different or if they meant nothing at all. (Dretzske, 1989, pp. 1–2).

Sosa, Honderich, Kim, and Dretske (among others)[8] think that, once we have made the distinction between the causally relevant and irrelevant features of an event, we will see that it is a consequence of AM that mental features are never causally relevant. Why is the causal irrelevance of the mental supposed to be entailed by AM? Kim (1984, p. 267) reasons as follows:

Consider Davidson's account: whether or not a given event had a mental description [...] seems entirely irrelevant to what causal relations it enters into. Its causal powers are wholly determined by the physical description or characteristic that holds for it; for it is under its physical description that it may be subsumed under a causal law. And Davidson explicitly denies any possibility of a nomological connection between an event's mental description and its physical description that could bring the mental into the causal picture.

The argument is that, since, according to AM, c causes e only if there is a strict law that subsumes c and e, and since strict laws contain only physical (never mental) predicates, it follows that the mental features of events c and e are irrelevant to whether the events are causally connected. The physical features of events suffice to fix, given the strict laws, *all* causal connections. Mental features neither suffice nor are required to fix causal connections. The argument is powerful. The conclusion which Kim draws from it is that, on AM, *the mind does not matter;* that a neural event has a certain intentional content is as irrelevant to its effect as the fact that the sounds are meaningful is to the sounds causing the glass to break.

But is this criticism of AM correct? We claim that it is not, and that it rests on a simple, but perhaps not obvious, confusion. The confusion is between two ways in which properties of an event c may be said to be causally relevant and irrelevant. Consider the following locutions:

(a) properties F and G are relevant$_1$ to making it the case that c causes e; and
(b) c's possessing property F is causally relevant$_2$ to e's possessing property G.

---

[8] Others who have argued that AM is epiphenomenalist include Stoutland, 1976, p. 307; Føllesdal, 1986, p. 315; Johnston, 1986, p. 423; and Skillen, 1984, p. 520.

We will say that (a) holds if and only if c has F and e has G and there is a strict law that entails that Fs cause Gs. It is in this sense that c's having F and e's having G 'make it the case' that c causes e. Relevance$_2$ is a relation among c, one of its properties F, e, and one of its properties G. It holds when c's being F brings it about that e is G. We shall argue that those who charge AM with epiphenomenalism are guilty of confusing relevance$_1$ with relevance$_2$.

None of the authors we have been considering defines the sense of causal relevance they have in mind when they accuse AM of rendering the mental causally inefficacious. Their discussions, though, do suggest a *test* for causal irrelevance. Recall Sosa's remark that 'had the gun been equipped with a silencer it would have killed the victim just the same' (Sosa, 1984, p. 278); and Dretske's remark that 'the glass would shatter if the sounds meant something completely different' (1989). So it may be that Sosa and Dretske (and others) think that AM entails the causal irrelevance of the mental, because they think that it entails the falsity of such mentalistic counterfactuals as 'if Fred had not believed that Jerry would attend the conference, he would not have come.'

In view of this counterfactual test for causal irrelevance$_2$, we suggest that the authors who propose it may have in mind the following characterization of causal relevance$_2$.[9]

[I] c's being F is causally relevant$_2$ to e's being G iff

(i) c causes e

(ii) Fc and Ge

(iii) $-$Fc $-$Ge

(iv) Fc and Ge are logically and metaphysically independent.[10]

Condition (iv) is intended to exclude cases in which the connection between F and G is conceptual/metaphysical rather than causal, for example, c's being the cause of e is causally relevant$_2$ to e's being caused by c when c does cause e.

The heart of our response to the claim that AM is committed to epiphenomenalism is this: AM entails that mental features are causally irrelevant$_1$, but it does not entail that they are causally irrelevant$_2$. Before arguing these claims, we need to discuss the interpretation of the counterfactual:

(Q) If event c were not F, then event e would not be G.

---

[9] While many philosophers appeal to the notion of causal relevance, it is far from clear that there is a single, or a well-characterizable notion that underlies the locution that c qua F causes e to be G. We are here interested only in sketching enough of an account to refute the charge that AM is committed to epiphenomenalism. Anyone interested in a thorough explication of causal relevance would have to show how to accommodate familiar difficulties involving pre-emption, overdetermination, and so on. But these are problems which confront every account of causation, and we will not discuss them here.

[10] c's being F and e's being G are metaphysically independent, iff there is a possible world in which c (or a counterpart of c) is F but e (or a counterpart of e) fails to occur or fails to be G and vice versa.

We will adopt the Lewis–Stalnaker[11] account of counterfactuals, according to which A > B is true if and only if B is true in all the worlds most similar to the actual world in which A is true (or A is true at no such world). We will suppose that an event e that occurs at the actual world may occur or have counterparts that occur at others. 'c' and 'e' are to be understood as rigid designators of events. In evaluating (Q), we need to look at the worlds most similar to the actual world in which c fails to be F. c may fail to be F at w either by existing there and not being F or by failing to occur at w (or have a counterpart) at all. (Q) is true just in case the most similar worlds at which counterparts to c fail to have F or at which c fails to have a counterpart are such that counterparts to e fail to have G or e fails to have a counterpart.

The irrelevance$_1$ of the mental follows immediately from the definition of relevance$_1$ and from AM's (4) and (6). The irrelevance$_1$ of psychological predicates, however, is perfectly compatible with the truth of counterfactuals $-Fc > -Ge$, where F and G are predicates that do not occur in strict laws. That is, the set of strict laws and basic physical facts does not by itself settle the truth values of counterfactuals.

We can see that this is so as follows: consider the set of worlds W at which all the strict laws hold. (This set includes the actual world a.) Until a *similarity order*, $\geq a$, is placed on W, the truth values of almost all counterfactuals are indeterminate. Only those counterfactuals A > B such that the strict laws and non-counterfactual statements true at a entail A → B or $-$(A → B) have determinate truth values, since any similarity ordering $\geq a$ will make the former true and the latter false. This is just the lesson of Nelson Goodman's (1982) failed attempts to analyze counterfactuals in terms of laws. What Goodman found is that laws and non-counterfactual truths are themselves not sufficient to settle the truth value of any but a limited set of counterfactuals. It follows that the truth of counterfactuals of the sort needed to establish causal relevance$_2$ (since neither they nor their negations are entailed by the strict physical laws and non-counterfactual truths) is compatible with AM.

Of course, it is one thing to show that mentalistic counterfactuals are compatible with AM. It is quite another thing to produce an account of what makes these counterfactuals true, and also to show that this account is compatible with AM. The question of what makes counterfactuals true is a general one, which concerns all counterfactuals and not just mentalistic ones. We shall briefly address it toward the end of our discussion.

To this point, we have shown that, if [I] supplies sufficient conditions for causal relevance$_2$, then there is no incompatibility between AM and the causal relevance$_2$ of the mental.[12] This is important, since, as we have seen, many of Davidson's critics seem

---

[11] See Lewis, 1973, and Stalnaker, 1968. There are differences between the two accounts, but they are irrelevant to our discussion.

[12] Although there is a tradition in the philosophy of action arguing that there are conceptual connections between propositional attitudes and actions, this does not entail that particular propositional attitude properties are conceptually connected. For example, suppose that John believes that Mary is across the street and, for this reason, he waves his hand. Let c be John's thought, e his action, F the property of his believing Mary is

to think that there is such an incompatibility. There are two further related questions we need to address. One is whether causal irrelevance$_1$ alone is sufficient to sustain a charge of epiphenomenalism. A second question is whether there are some further conditions on [I] such that, once they are added, AM does entail the causal irrelevance$_2$ of the mental.

Why would anyone think that irrelevance$_1$ of the mental entails epiphenomenalism? Honderich (1982) formulates a principle which he calls 'the principle of the nomological character of causally relevant properties', according to which c's having F is causally relevant to e's having G, if and only if there is a law of the form Fs cause Gs (p. 62). If one thinks, as Honderich does, that AM implies that psychological predicates never appear in causal laws, then one might conclude that psychological features have no causal role to play, and indeed that psychology could not be a science. But, as Davidson has been careful to observe (1980, p. 240), there may very well be psychological and psychophysical causal laws that support counterfactuals and other subjunctive conditionals; it is just that such laws cannot be *strict*. If Honderich intends for the principle of nomological relevance to include non-strict as well as strict laws, then AM is compatible with the causal relevance (in Honderich's sense) of psychological properties. If he intends for the principle to include only strict laws, then the principle is unacceptable. It is implausible that there are any strict laws linking 'is a match striking' with 'is a match lighting'. So, on the strict law construal of Honderich's principle, being a match striking is not causally relevant to the match's lighting. On this construal, Honderich's principle would render virtually all properties of events causally irrelevant$_2$. This certainly seems wrong.

In arguing that AM entails the causal irrelevance of the mental, some authors have suggested a strengthened account of causal relevance$_2$. For example, Sosa writes:

I extend my hand because of a certain neurological event. That event is my sudden desire to quench my thirst. Thus, if my grasping is caused by that neurological event, it's my sudden desire that causes my grasping [. . .] Assuming the anomalism of the mental, though extending my hand is, in a certain sense, caused by my sudden desire to quench my thirst, it is not caused by my desire qua desire but only by desire qua *neurological* of a certain sort. [. . .] [T]he being a desire of my desire has no causal relevance to my extending my hand (if the mental is indeed anomalous): *if the event that is in fact my desire had not been my desire but had remained a neurological event of a certain sort, then it would have caused my extending my hand just the same.* (Sosa, 1984, 277–278, our emphasis.)

This passage suggests the following as a sufficient condition for causal irrelevance$_2$:

[II] c's being F is causally irrelevant$_2$ to e's being G, if there is a property F* of c such that (F* c & $-Fc$) > Ge holds non-vacuously.

across the street, and G the property of being a waving hand. Clearly we can have c's being F causally relevant$_2$ to e's being G, since c's being F can obtain without e's being G and vice versa in some metaphysically possible world.

Even when $-Fc > -Ge$ holds, there may be a property $F\star$ of c such that $(F\star c \ \& \ -Fc) > Ge$. In this case, it may seem that it is in virtue of c's being $F\star$, not F, that e is G. When this holds, we will say that $F\star c$ 'screens off' Fc from Ge. Converting [II] into a necessary condition for causal relevance$_2$ and adding it to [I], we obtain the following proposal:

[III] c's being F is relevant$_2$ to e's being G iff the conditions in (I) are satisfied and there is *no* property $F\star$ of c such that $(F\star c \ \& \ -Fc) > Ge$ holds nonvacuously.

Sosa seems to think that it follows from AM that c's being a certain neural state, Nc, *screens off* c's being a desire to quench thirst, Mc, from e's being an extending of the hand, Be. More generally, he seems to think that neural properties screen off intentional mental properties. Presumably Sosa thinks that this follows from AM, because he thinks there are strict laws connecting neural properties with behavioral properties. Since mental properties are not reducible to neural proper- ties, it follows that there are physically possible worlds in which Nc, Mc, and in *all* such worlds Be.

It is not at all clear that there are strict laws connecting neural properties with mental properties (and so that AM *entails* that the neural property screens off the mental property), but it does seem that, as a matter of fact, in a case like Sosa's, the neural property does screen off the mental property. The worry then is that, if [II] is kept as a condition on causal irrelevance$_2$, then the causal irrelevance$_2$ of the mental will follow from AM after all.[13]

In response to this, notice first that [II]'s rendering the mental causally irrelevant$_2$ is independent of AM, at least to the extent that the problem-creating counterfac- tual, (Nc & $-$Mc) > Be, holds whether or not there is a strict law linking N with B. So anyone who adopts [II] as a condition on causal irrelevance$_2$ will be committed to the causal irrelevance$_2$ of the mental in this case. But it seems to us that [II] is not a correct condition on irrelevance$_2$. It renders even properties connected by strict law causally irrelevant$_2$. To see this, consider the neural event c and the behavioral event e in Sosa's example. c possesses the basic physical property P and the mental property M (being a desire to quench one's thirst), and e possesses the property B (being a certain movement of the hand). Assuming a strict law between P and B, it follows that

---

[13] Jerry Fodor (1987, Ch. 2) has argued that a taxonomy of propositional attitude states in terms of their truth conditions is not a taxonomy in terms of causal powers. Condition [III] may be involved in the view of some philosophers that scientific psychology requires a notion of narrow content. Thus, Fodor seems to hold that Oscar's belief that water quenches thirst is not causally relevant $_2$ to Oscar's behavior, since, if Oscar were in the same neural state as he is in but had not believed that water quenches his thirst, he would have behaved identically. The antecedent of this counterfactual is thought to be metaphysically possible for Putnamian reasons: if Oscar has lived in an environment containing XYZ and not H$_2$O, his neural state would have been a belief that twin water quenches thirst. One might conclude that, if we want a notion of content such that propositional attitudes are causally relevant$_2$ in virtue of their contents, then we need a notion of content which makes propositional attitudes supervene on neural states.

(S) $(-Mc \& Pc) > Be$.

So P screens off M from B. Now consider the counterfactual

(T) $(-Pc \& Mc) > Be$.

It can be shown that (T) is compatible with AM *and* with (S). Furthermore, it is plausible that (T) is in fact true. If c had been a desire to quench thirst but had not been P, it would have had some other property $P^\star$. Furthermore, c would still have resulted in an e that has the property B. That is, in the closest possible world in which Sosa desires to quench his thirst but this desire is not a P, the desire still causes him to extend his hand. Supporting this claim there may be a law, though not strict, to the effect that, when someone experiences a sudden desire to quench her thirst and believes that there is a glass of water in front of her which she can reach by extending his hand, then, *ceteris paribus*, her hand will extend. When we consider the possibility that c is M but not P, this law 'takes over', so that c still causes an event that is B. Here is a non-psychological example which will, perhaps, help elucidate our claim.

Consider the event of Hurricane Donald striking the coast and causing the streets to be flooded. That event is identical to the event of certain air and water molecules moving in various complex ways. Call the property of consisting of molecules moving in such ways P. It is perfectly possible for the following counterfactual to be true: if hurricane Donald had not had property P (that is, if a hurricane as much like Donald as possible, though without P, had occurred), then it would still have caused the streets to be flooded. Indeed, it would have had some property $P^\star$ sufficiently similar to P, and $P^\star$ events (under the relevant conditions) cause floodings. The result is that Donald's being a hurricane would be said to be causally irrelevant to its flooding the streets. We think that examples such as this one show that [III] is too strong a requirement on causal relevance$_2$.[14]

A fully adequate account of causal relevance$_2$ should show how mentalistic counterfactuals are grounded. What is it about Sosa, his situation, and so on, that makes it true that, if he had not experienced a sudden desire to quench his thirst, he would not have extended his hand? We do not have such an account, but we want to suggest an approach that fits within the framework of AM. As we have observed already, the existence of non-strict psychophysical and psychological laws is compatible with AM. A non-strict law is one which has a *ceteris paribus* qualifier. The interesting thing about such laws is the ways in which they can support counterfactuals. We will illustrate this by building upon a suggestion made by Lewis (1975). Let R, W, and B be the statements that a red block, a white block, and a blue block are placed in front of Donald, and let S_r, S_w, and S_b be the statements that Donald sees a red block, a white

---

[14] It may be that there is some account of causal relevance$_2$ midway in strength between [I] and [III], which captures what some of what Davidson's critics have in mind. We leave them the task of formulating it and attempting to demonstrate that AM entails the irrelevance$_2$ of the mental so characterized.

block, and a blue block. We will suppose, as is plausible, that there are non-strict laws of the form:

(L) If X and C, then S_x,

where C are conditions like 'lighting is good', 'Donald is awake and paying attention', and so on. Even with such conditions added, the law is a *ceteris paribus* one and, if AM is correct, it will be impossible to add explicit conditions that turn it into a strict law. When the laws (L) hold, we will say that the statements describing what Donald sees *depend nomically* on the statements describing the blocks in front of him. Call conditions C *counterfactually independent* of the family of statements {R, W, B}, if C would continue to hold no matter which member of {R, W, B} is true. Lewis shows that, if C and the *ceteris paribus* conditions associated with (L) are counterfactually independent of {R, W, B}, then S_x will depend counterfactually on X. That is, each of the counterfactuals, $R > S_r$, $W > S_w$, $B > S_b$, will be true. If we further assume that a block which has one of three colors will be placed in front of Donald (and that this statement is also counterfactually independent of (R, W, B)), then the statement $-X > -S_x$ will also be true. Suppose that a red block is placed in front of Donald, and that this event causes the event of his seeing a red block. It will follow that, if the first event had not been a placing of a red block, then the second event would not have been Donald's seeing a red block. As Lewis points out, this 'grounding' of counterfactuals in laws fails to *reduce* counterfactuals to laws, since the assumption of counterfactual independence is essential. It does show, however, how laws, including *ceteris paribus* laws, can support counterfactuals. The program for a psychology compatible with AM is the discovery and the systematization of such non-strict laws (at various levels) connecting psychological and/or behavioral properties.

We have seen that AM attempts to resolve the mind–body problem by endorsing (2), (3), and (3′), by denying (1) with respect to events, and by affirming (1) with respect to properties. Davidson is silent on (2) and (3) with respect to properties, which leads to the accusation that AM is committed to epiphenomenalism. We rebutted this charge by showing that AM is compatible with there being counterfactual dependencies between events in virtue of their mental properties. To do this is to affirm (2) with respect to properties but, of course, to deny (3) with respect to properties. An event's physical features may counterfactually depend on another event's mental features. But, interestingly, we need not deny (3′) for our account of causal relevance$_2$. It may be that all counterfactuals *supervene* on basic physical truths and strict laws. That is, if two possible worlds are exactly alike with respect to basic physical facts and strict laws, they are exactly alike with respect to counterfactuals. This fairly strong physicalism still allows sufficient autonomy of the mental, so that it is not reducible to the physical and it has a genuine explanatory and causal role to play.

Acknowledgments

A former version of this chapter was presented in an APA symposium of the same title, held on December 30, 1987. Thanks are due to Jonathan Adler, John Biro, Paul Boghossian, Donald Davidson, Fred Dretske, Ray Elugardo, Jerry Fodor, Richard Foley, Terry Horgan, Brian McLaughlin, Alexander Rosenberg, Stephen Schiffer, and John Searle.

References

Achenstein, P. (1977) The Causal Relation. *Midwest Studies in Philosophy* 5: 369–386.

Davidson, D. (1980) *Essays on Actions and Events.* New York: Oxford University Press.

Dretzske, F. (1989) Reasons and Causes. *Philosophical Perspectives* 3: 1–15.

Fodor, J. (1987) *Psychosemantics.* Cambridge, MA: MIT Press.

Føllesdal, D. (1986) Causation and Explanation: A Problem in Davidson's View on Action and Mind. In E. Lepore and B. McLaughlin (eds), *Actions and Events: Perspectives on the Philosophy of Donald Davidson,* London: Basil Blackwell, pp. 311–323.

Goodman, N (1982) *Fact, Fiction and Forecast,* 4th edn. Cambridge, MA: Harvard University Press.

Honderich, T. (1982) The Argument for Anomalous Monism. *Analysis* 42/1: 59–64.

Honderich, T. (1984) Smith and the Champion of the Mauve. *Analysis* 44/2: 86–89.

Johnston, M. (1986) Why the Mind Matters. In E. Lepore and B. McLaughlin (eds), *Actions and Events: Perspectives on the Philosophy of Donald Davidson,* London: Basil Blackwell, pp. 408–426.

Kim, J. (1984) Epiphenomenal and Supervenient Causation. *Midwest Studies in Philosophy* 9: 257–270.

Kim, J. (1986) Psychophysical Laws. In E. Lepore and B. McLaughlin (eds), *Actions and Events: Perspectives on the Philosophy of Donald Davidson,* London: Basil Blackwell, pp. 369–386.

Lewis, D. (1973) *Counterfactuals.* Oxford: Blackwell.

Lewis, D. (1975) Causation. In E. Sosa (ed.), *Causation and Conditionals,* Oxford: Oxford University Press, pp. 180–191.

Mackie, J. L. (1979) Mind, Brain and Causation. *Midwest Studies in Philosophy* 6: 19–29.

McLaughlin, B (1986) Introduction. In In E. Lepore and B. McLaughlin (eds), *Actions and Events: Perspectives on the Philosophy of Donald Davidson,* London: Basil Blackwell, pp. 3–13.

Searle, J. (1983) *Intentionality.* Cambridge: Cambridge University Press.

Skillen, A. (1984) Mind and Matter: A Problem that Refuses Dissolution. *Mind* 93: 514–526.

Sosa, E. (1984) Mind–body Interaction and Supervenient Causation. *Midwest Studies in Philosophy* 9: 271–282.

Stalnaker, R (1968) A Theory of Conditionals. In N. Rescher (ed.), *Studies in Logical Theory,* Oxford: Basil Blackwell, pp. 98–112.

Stoutland, F. (1976) The Causation of Behavior. In Acta Philosophica Fennica: *Essays on Wittgenstein* in Honour of G. H. von Wright, 28, pp. 286–325.

12

More on Making Mind Matter

Ernest Lepore and Barry Loewer

During the heyday of neo-Wittgensteinian and Rylean philosophy of mind, the era of the little red books, it was said that propositional attitude explanations are not causal explanations and that beliefs, intendings, imaginings, and the like are not even candidates to be causes. Indeed, to treat mentalistic language as describing causes or causal processes is, it was said, to commit a logical error. We have come a long way since then. The work of Davidson, Armstrong, Putnam, and Fodor (among others) has reversed what was once orthodoxy, and it is now widely agreed that propositional attitude attributions describe states and episodes which enter into causal relations. But, as a number of writers have recently observed, an F may cause a G even though its being an F is irrelevant to its causing a G.[1] For example, Fred's taking a round pill may cause him to fall asleep, but the shape of the pill is irrelevant (we may suppose) to its causing sleep. Similarly, Donald's throwing a green stone with a certain force may cause the window to break. That his throwing is of a *green* stone is irrelevant, while its being *with a certain force* is relevant to it causing the window's breaking. It seems that an event's properties are not all on an equal footing when it comes to causation.

How should we characterize 'causal relevance'? We will not attempt to answer this question yet but we will make a few preliminary points. First, c's possessing F may be relevant either to c's causing e or to c's causing a G. In the first case, c's having F in some way (yet to be described) *grounds* c's causing e. In the second case, c's having F grounds its causing a G. We think of the grounding relation as both metaphysical and explanatory. It is metaphysical since it involves some real relation between F and c and e (or G). It is explanatory since citing this property and the relation will, at least under certain circumstances, explain why c caused e (a G). We will say that a property P is *causally potent* just in case it is capable of grounding some causal relations. The idea is that it is the force with which the rock was thrown, and not its being a green rock, that grounds the causal relation between the throwing and the breaking. Of course, the throwing

---

[1] The distinction between properties which are and those which are not relevant to an event c's causal relations has been made by a number of authors; for references, see the previous chapter. The first of the examples is from Dretske, 1989.

of a green rock may be causally potent, since it might ground some other causal relation.

Even without an explicit account of causal relevance, a number of philosophers of mind have lately begun to worry that content properties may be as causally irrelevant to the effects of events which instantiate them, as the pill's shape is to its dormitive powers. This worry arises from the recognition that content properties possess certain features which apparently make it impossible to identify them with neurophysiological or other properties (e.g. syntactic), whose causal relevance is thought to be unproblematic. We will later examine some of the reasons why the non-identity of content and neurophysiological properties should be thought to render the former causally impotent. For now we note that, if content properties are causally irrelevant, then we are faced with a new kind of epiphenomenalism: 'content property epiphenomenalism.' According to content property epiphenomenalism, while events which instantiate content properties may enter into causal relations, their content properties themselves have no bearing on these causal relations.

One is likely to think that content property epiphenomenalism is contrary to common sense. It certainly seems that, when Clyde's reason for going to the fridge is his thought that there is beer in the fridge, the thought's possessing its content—that there is beer in the fridge—has something to do with the fact that it causes Clyde to go to the fridge. Furthermore, actions are not the only consequences of thoughts to which their content properties seem causally relevant. The content of Clyde's thought about beer makes a difference not only to his actions but also to the states of things which are not described intentionally, for instance to the location of Clyde's body. And wondering whether there are infinitely many primes will, other things equal, initiate a sequence of thoughts dependent on the content of that wondering. Indeed, it would be surprising and dismaying to learn that discoveries about the nature of content establish that the content properties of Clyde's thoughts have no causal role to play with respect to his behaviour.

In this chapter we will be concerned with the question of what, if any, causal role content properties play. In the first part, we will discuss considerations which may lead one to worry that content properties are causally impotent. These worries arise from combining certain physicalistic commitments with the view that content properties are not reducible to physical properties. So in the first part, we will formulate a non-reductive physicalism and then canvass some reasons for thinking that content properties are not reducible to physical basic properties. In the second part, we will consider three attempts to characterize a causal for content and we will show how content properties can play these roles.

Non-reductive physicalism?

The primary reason why Descartes' dualism of mental and physical substances and events seems to us to be so unsatisfactory is that it is difficult to square it with any reasonable account of mental causation. Not only is it difficult to understand how a non-extended mental substance can interact causally with something physical, but the advances of the sciences have made it increasingly plausible that the physical realm is causally closed. By this we mean that the occurrence of any physical event can be causally explained (to the extent that it is possible to explain it) by invoking only other physical events and laws. An interactive event dualism obviously violates the closed nature of the physical. The token identity theory, in contrast to event dualism, is able to countenance mental events as causes without infringing on the closed character of physical causation since it identifies mental events with physical events But token physicalism is a fairly tepid physicalism. It is compatible with there being two worlds, which would be exactly alike in their physical laws and facts and yet differ in other respects. It is even compatible with emergent causation, the view that there are causal relations among events which are not determined by those events' physical properties. A stronger physicalism is embodied in the following three principles:

(i)]sh[*The comprehensiveness of physics.* There exists a fundamental theory (or theories) of basic physics such that every event instantiates a causal property of the theory(or theories). It follows that every event is a physical event.

(ii)]sh[*Global supervenience.* If two nomologically possible worlds are exactly alike with respect to fundamental physical facts (that is, to the facts expressible in terms of the vocabularies of fundamental physical theories), then they are exactly alike with respect to all other facts.

(iii)]sh[*The priority and nomologicality of physical causation.* For any events c and e, c causes e if and only if there is a causal law of the fundamental theory of physics which subsumes c and e.

We will call the metaphysical view embodied in (i) through to (iii) *non-reductive physicalism.*[2] This view requires considerable elaboration and defense, but we will limit ourselves here to just a few remarks. First, we have to admit that we have no definition to offer of 'fundamental physical theory'. This will pose no problem for our discussion, since one generally has a good idea as to which theories of our world are likely to be fundamental physical theories and which theories are not; for instance quantum theory is a fundamental physical theory, while price theory is not. A property is a basic physical property if it is expressed by a predicate of a fundamental physical theory and the basic physical facts are the facts expressed by statements of a fundamental physical theory. In (i) we are assuming that the fundamental physical theories are comprehensive in that they possess the resources for specifying every event (though, of course, we might not know how to redescribe a particular event, say, the assassination of the Archduke, in terms of the fundamental physical theory).

---

[2] The similarity between non-reductive physicalism and Davidson's anomalous monism is of course intentional. We discuss the differences in Chapter 11 above.

Global supervenience expresses the idea that all facts depend on physical facts. Two nomologically possible worlds which agree on physical facts also agree on mental facts, counterfactuals, and so forth. But this dependence is of a fairly weak sort. In particular, the global supervenience of all facts on physical facts does not imply that every property is, or is reducible to, a basic physical property, or even that there are one-way bridge laws of the form $P \rightarrow M$ connecting basic physical properties with, for instance, mental properties. The reason for the latter is that bridge laws involve *explanatory* connections between properties, and there is no guarantee that the global necessary connections entailed by global supervenience will be explanatory.

Condition (iii) is a special case of global supervenience, since it implies that causal relations are determined by physical facts. It is stronger than global supervenience, because it asserts that the basic physical properties of events c and e and the basic physical laws determine whether or not there is a causal relation between c and e. Thus the fundamental causal laws and the distribution of causal physical properties at a world completely determine the causal relations which obtain at that world. This doesn't mean that there are no laws couched in terms of predicates which express non-basic properties. It is compatible with non-reductive physicalism that there exists such laws, even causal laws. Such laws may enjoy a certain autonomy with respect to the laws of basic physics, since it may be impossible to reduce them to fundamental physical laws. The status of such laws in *grounding* causal relations will be discussed later.

How does causal relevance fit into non-reductive physicalism? If c is P and e is P* and *Ps cause P*s* is a fundamental physical law, then P is causally relevant to c's causing e. This seems correct, since c's being P accounts for it causing a P*. Can properties other than P, and particularly properties which are not reducible to physical properties, also be causally relevant to c's causing e? We will return to this question after considering some of the reasons that have recently been offered to support the irreducibility of content properties.

The irreducibility of content properties?

Recent discussions of content properties have focused on a number of features which these properties are claimed to possess and which have been thought to show either that they are not reducible to the physical properties which ground causal relations (i.e. to physical properties which occur in causal laws) or not reducible to physical properties at all. These features are multiple realizability, non-supervenience on neurophysiological properties, and normativity. We want to review briefly some of these features and to explain why each may lead one to worry that content properties are epiphenomenal.

It is practically received wisdom among philosophers of mind that psychological properties (including content properties) are not identical to neurophysiological or other physical properties. The relationship between psychological and neurophysiological properties is that the latter *realize* the former. Furthermore, a single psychological property might (in the sense of conceptual possibility) be realized by a large perhaps infinitely many, number, of different physical properties, and even by non-physical properties. (See the articles on functionalism in Block, 1980.)

Exactly what is it for one property of an event to *realize* another? The usual conception is that e's being P realizes e's being F iff e is P and e is F and there is a strong connection of some sort between P and F. We propose to understand this connection as a necessary connection which is *explanatory*. The existence of an explanatory connection between two properties is stronger than the claim that P → M is physically necessary, since not every physically necessary connection is explanatory. For e's being P to explain its being F it may be necessary for there to be a *system* of connections between realized and realizing properties of property kinds to which P and F belong. And it may require that the central laws and principles governing the realized properties be explained by the connections between basic and non-basic properties and laws governing the basic properties. For example, systematic variations in the molecular structure of a substance give rise to systematic variations in its degree of solubility, and possession of a particular molecular structure explains the rate of dissolving and so forth. This supports the claim that a substances molecular structure realizes its solubility.

If there are infinitely many physical (and perhaps non-physical) properties which can realize F, then F will not be reducible to a basic physical property. Even if F can only be realized by finitely many basic physical properties, it might still not be reducible to a basic physical property, since the disjunction of these properties might not itself be a basic physical property (i.e. occur in a fundamental physical law). We will understand 'multiple realizability' as involving such irreducibility.

To see how the view that content properties are multiply realizable by neurophysiological (and other physical) properties can give rise to worries about the causal potency of content properties, consider the following diagram:

In the diagram event c, a certain pattern of neuron firing, causes event e, a certain motion of the hand. P_1 is a neurophysiological property, P_2 is a certain movement-of-the-hand-property, M_1 is the property of thinking 'that's my friend across the street' and M_2 is the property of waving. We suppose that P_1 realizes M_1 in c and P_2 realizes M_2 in e. Finally, suppose that c's being P_1 is sufficient for it to cause e and to cause e to be P_2.

Let us suppose that P_1's *cause* P_2's is a fundamental causal law. In that case P_1 is causally relevant to the causal transaction between c and e. The question is whether c's being M_1 is causally relevant as well. To be causally potent, it would have to explain or ground the causal relation between c and e or account for why the event caused by c is M_2. The epiphenomenalist worry is that it does neither. In support of this, the epiphenomenalist can cite that c's being P_1 is sufficient to account for its causing e and for e's being P_2 and, given the fact that e's having P_2 explains its possession of M_2, it is also sufficient to account for M_2. It seems that c's having M_1 is superfluous. It is not needed to ground the causal or explanatory relations in which c is involved, since c's neurophysiological properties already do that. Notice that *multiple* realizability gives rise to the problem, since, if M_1 was realized only by P_1, then it would be possible to identify M_1 with P_1 and there would be no problem about the causal relevance of M_1 (over and above any problem about the causal relevance of P_1).

A second feature of content properties is that they fail to supervene on neurophysiological properties. This means that two individuals may be exactly alike with respect to their neurophysiological (or basic physical) properties and yet differ with respect to their content properties. For example one may believe that there is beer in the fridge and the other that there is 'twin beer' in the fridge. Supervenience failure is taken to be a consequence of Putnam's famous Twin Earth thought experiment, which presupposes a causal account of natural kind terms (we spare the reader another recitation of Putnam's stories). But it is also a consequence of any externalist account of content. It follows from non-supervenience not only that no neurophysiological property is identical to the psychological property of believing that there is beer in the fridge (as multiple realizability entails), but also that no neurophysiological property even necessitates the psychological property. This doesn't mean that *no* physical property realizes contentful psychological properties. There may be a more global physical property, whose instantiation explains the possession of content properties.[3]

Failure to supervene on neurophysiological properties has been taken to disqualify a property from being causally potent for the following reason. We may suppose that there is a sufficient (or as close to being sufficient as is allowable by physical theory) causal explanation of a person's movement in terms of his prior neurophysiological state (and local environmental conditions in the vicinity of his body, e.g. whether he is in water or in the gravitational field). If this is so, then he would have moved exactly the same way even if he had a different belief, as long as his neurophysiological properties remained the same (e.g. if his belief was about twin water rather than water). The antecedent of this counterfactual is not vacuous since, due to failure of supervenience, there are possible worlds in which the neurophysiological state is the same but the content of the psychological state is different. The conclusion is that the

---

[3] An externalist account of content is one which claims that content properties are constituted (in part) by relations between the bearers of content and something external, e.g. causes of tokening of the content. Davidson, Dretske, Burge, and Fodor all hold (different) externalist accounts.

content property *believing that there is water over there* is superfluous and hence is not causally potent, at least with respect to a person's movements.

Jerry Fodor has produced a related argument to a similar conclusion. He argues that content properties cannot affect what he calls 'the causal power' of events that possess them. By 'causal power of an event' he means the event's ability to enter into causal relations with other events and states of affairs. The argument is as follows (see Fodor, 1987, p. 39).

A property P can affect x's causal powers only if there is a causal mechanism or a fundamental causal law connecting x's having P with x's causal powers. But there is no causal mechanism or basic causal law that connects x's being a belief that there is water in the glass with its causal powers. The reason is that 'you can't affect the causal powers of a person's mental states without affecting his physiology' (ibid.). But, presumably, because of the failure of content properties to supervene on neurophysiological properties, you can affect the former without affecting the latter. So x's being a belief that p is not a property that affects causal powers. The reason that there is no such causal mechanism seems to be that the contents of x's beliefs are determined by causal relations between her environment and her brain, or by the deference conventions in her speech community, and these can be changed while leaving her physiology the same.

The third feature opens the widest gap between content properties and physical properties. It has been argued by a number of philosophers that content properties are not physicalistically explicable. The underlying reason is that content properties are essentially normative properties. The normative aspects at issue arise from the possession of truth conditions. In virtue of possessing their truth conditions, contentful states are *essentially* truth evaluable and essentially evaluable in terms of their logical and rationalizing connections to other thoughts and attitudes. That is, to be a state with content is to be a state which is assessable as true or false, as rational or irrational.[4] If this is correct, then content properties are not even *realized* by physical properties—in the sense of 'realize' discussed above.

This kind of property dualism is compatible with psychological facts globally supervening on physical facts, and even with the existence of necessary connections of the sort N (P→M) (where P is a physical property and M is a content property). But, if there are such necessary connections, they are not sufficiently systematic and lawful to provide an explication of the content properties. In particular, the connections are incapable of accounting for the normative features of content properties.

We will not discuss whether or not content properties are physicalistically explicable. But we do want to mention a reason for thinking that, if they are not, then they are

---

[4] There are two senses in which we might speak of properties or predicates as being normative. Predicates like 'is good' and 'is true' are normative in that they are employed in making normative assessments. A predicate like 'has the content that there is beer in the fridge' is normative in that something which satisfies it is, in virtue of satisfying it, normatively assessable, in this case in terms of truth value.

causally impotent.[5] Suppose one held that, for a non-basic law like *Fs cause Gs* to be a genuine causal law capable of grounding causal relations, there should be more basic explanations for why it holds. Without such explanations its truth would be merely coincidental, and so the law would not be a genuine law. The natural way of explaining why *Fs cause Gs* holds of *c* and *e* would be to locate the properties P and P* of *c* and e which realize F and G. Then one would have an account of how and why this F causes a G. Of course, if there is multiple realizability, then we would need different explanations in different cases. But, if F is not realized by any physical property, then there would be no explanation of this sort for why *Fs cause Gs holds*. If we suppose that only a property which occurs in a genuine causal law is causally potent, then it follows that content properties are not causally potent.

The above considerations all favour the view that content properties are not reducible to the physical properties which occur in fundamental physical causal laws. The extent and nature of the irreducibility varies among the considerations. If F is multiply realizable (but supervenient on causal physical properties) and is realized by P, there is a rather tight connection between F and P. Perhaps it is sufficiently tight to attribute causal efficacy to P. The normativity considerations, on the other hand, open up a wide gap between content properties and physical properties. For now we want to abstract from the various reasons for irreducibility and focus on an argument which proceeds from irreducibility to the causal impotency of content properties:

(1) Content properties are not reducible to neurophysiological or other properties which occur in the causal laws of fundamental physical theories (since they are multiply realizable, non-supervenient on neurophysiological properties, and normative).

(2) *c*'s causal relations are determined by those of *c*'s properties which occur in the causal laws of fundamental physical theories or properties which are reducible to them.

(3) So any property of *c* which does not occur in a fundamental causal law or is not reducible to such properties does not determine *c*'s causal relations.

(4) A property of *c* is causally potent only if it determines some of *c*'s causal relations.

(5) So content properties are causally impotent.

This argument propounds a paradox. The irreducibility of content properties, the priority of physical causation, and the causal potency of content properties are all plausible. But, according to the argument, they are incompatible. How should one respond to the paradox? The working assumptions of this chapter—non-reductive physicalism—validate premises (1) and (2). Proposition (3) is represented above as following from (2). In fact it doesn't. It is possible that, even though an event's physical

---

[5] Whether or not content properties are physicalistically or naturalistically explicable is a matter of considerable controversy. The most interesting proposals for naturalizing content are due to Dretske, 1981 and Fodor, 1987 and. Criticisms of these approaches are to be found in Loewer, 1987 and Boghossian (1990).

properties completely determine its causal relations, some of its other properties, including its content properties, also determine (some of) its causal relations. That is, some of an event's causal relations may be *overdetermined*. If this is so, then content properties are causally potent. A different response would be to reject (4). This would involve finding a construal of 'casual potency' which assigns causal work to a property even though that property does not determine an event's causal relations. In the next section we will examine three attempts to respond to this argument.

Kim's supervenient causation?

Jaegwon Kim has proposed that causal relations among macro-events, and specifically among mental events, are dependent on causal relations among basic physical events (Kim, 1983 and 1984). He calls such dependent causal connections 'epiphenomenal' or 'supervenient.' His use of the term 'epiphenomenal' emphasizes the dependent nature of these causal relations, while 'supervenience' suggests the nature of the dependence. Kim construes an event as an ordered triple consisting of an object, a time, and a property $\langle x,t,P \rangle$; the event of x's possessing P at t. We have been construing events in Davidsonian fashion as more coarse-grained. On this account of events, supervenient causation is best understood as a relation between (coarse-grained) event-property pairs. The characterization becomes:

$\langle c, F \rangle$ superveniently causes $\langle e, G \rangle$ iff there are basic properties P and P* such that N (P $\rightarrow$F) and N (P $\rightarrow$G) and c causes e in virtue of c's being P and e's being P*.

If we accept the nomological account of causation specified in (iii), then the final clause of the above definition can be replaced by 'c and e are subsumable under the basic causal law "Ps cause P*s."'

For example, the event c's being the heating of a piece of metal superveniently causes e's being the metals expanding, because the molecular property which realizes the heating is causally related in the appropriate way to the molecular property which realizes the expanding.

Can supervenient causation be used to resolve our paradox? If so, we must understand it as providing a characterization of causal potency. If it is to provide a successful resolution, it must satisfy the following conditions:

(a) it is a causal relation sufficiently strong to vindicate our commonsense views concerning content causation;
(b) content properties must enter into supervenient causal relations;
(c) their doing so must be compatible with physicalism.

Before discussing whether or not content properties enter into supervenient causal relations, we can see that non-basic properties which are multiply realizable can enter into such relations. While there must be a physical property P for which N (P $\rightarrow$ F) obtains, F may not be reducible to physical properties—either because it is multiply realizable or because the necessary relation is not an explanatory relation. So step (4) in our epiphenomenalist argument is false when causal potency is characterized in terms of supervenient causation. F may be superfluous to c's causal relations—these are already determined by c's physical properties—but still c's being F superveniently causes e's being G. Furthermore, the claim that $\langle c, F \rangle$ superveniently causes $\langle e, G \rangle$ is compatible with non-reductive physicalism. If c's being F superveniently causes e's being G, then not only must F and G supervene on basic physical facts, but the supervenient causal relation is completely dependent on the existence of a causal relation between the basic physical properties. Thus, condition (c) is satisfied.

Is supervenient causation a causal relation sufficiently strong to underwrite our commonsense views concerning content causation? Kim's strategy for supporting his claim that supervenient causation is causal enough to rebut epiphenomenalism is to argue that macro-properties like the heating of metal, which we ordinarily take to be causal properties, participate in supervenient causal relations. If psychological properties also participate in supervenient causal relations, then they are, at least in this respect, no worse off than macro-properties. Kim concludes that 'this seems sufficient to redeem the causal powers we ordinarily attribute to mental events' (1990, p. 268). So, although supervenient causation does not accord psychological properties the same causal potency as basic physical properties have, it does accord them, Kim claims, the same causal status that is enjoyed by non-psychological macro-properties.

For Kim's reasoning to be compelling, macro-properties like the heating of metal must possess their causal potency in virtue of being supervenient causes. But this is dubious. The supervenient causal relation is much too weak to ground a substantial notion of causal potency. Whatever it is that underlies our judgement that the instantiation of a macro-property (e.g. c's being heated) is causally potent, it cannot be supervenient causation. Brian McLaughlin (1983) has pointed out one way of seeing why this is the case. If c's being F superveniently causes e's being G and F$\star$ and G$\star$ are any properties for which N (F $\rightarrow$ F$\star$) and N (G $\rightarrow$ G$\star$), then c's being F$\star$ superveniently causes e's being G$\star$. For example, assuming that c's being the heating of a metal superveniently causes e's being its expanding, then, since heating entails increasing energy, then c's being an increasing of energy in the metal superveniently causes e's being an expanding of the metal. But this doesn't seem right. Not any kind of energy increase in the metal causes it to expand. So it appears that the instantiation of a macro-property can superveniently cause the instantiation of another macro-property without there being a 'real' causal relation between them.

What is present in the case of heating causing expanding that is not guaranteed by supervenient causation? There are two obvious suggestions. First, notice that, for a piece of copper under usual circumstances, the following counterfactuals will generally be true: if the piece of copper were heated, it would expand; if the piece of copper were not heated, it would not expand. But $\langle c, F \rangle$ may superveniently cause $\langle e, G \rangle$, even though neither of the corresponding counterfactuals is true. This point is telling against supervenient causation as an account of content causation, since in expressing anti-epiphenomenalist intuitions it is natural to appeal to counterfactuals involving content properties; say, if he had not thought that the beer was in the fridge, he would not have gone there.

There is a second feature present in the heating–expanding example that may be absent from supervenient causal relations in general. The fact that heating a piece of metal causes it to expand is a law. This law is, of course, not a fundamental law of physics, and is best construed as a *ceteris paribus* law. But it is nevertheless a law. $\langle c,F \rangle$ may superveniently cause $\langle e,G \rangle$ even though there is no law, not even a *ceteris paribus* law, that Fs cause Gs. Indeed, it may even be a *ceteris paribus* law that Fs cause –Gs.

The above considerations show that Kim's defense of supervenient causation as providing a 'substantial' notion of causal potency fails. The existence of supervenient causation relations involving content properties is not sufficient for rebutting epiphenomenalism. Furthermore, there are reasons for thinking that content properties cannot even enter into supervenient causation relations. Externalist considerations show that events c and c can be identical with respect to their neurophysiological and other causal properties, though they differ with respect to their content properties. If this is so, then c's content properties do not supervene on its causal properties and so they do not enter into supervenient causal relations.[6]

Our conclusion is that supervenient causation does not provide a solution to content property epiphenomenalism. It is not a sufficiently strong relation to allay epiphenomenalist fears, and it seems that content properties do not enter into supervenient causal relations.

Fodor's causal powers

Jerry Fodor (1990) defends the view that F is a *causally relevant* property just in case there is a causal law to the effect that Fs cause Gs. The same view is developed by Brian McLaughlin (1990) in the course of defending Davidson's anomalous monism against the charge that it is committed to property epiphenomenalism. We will understand these writers as attempting to characterize a notion of causal potency—we will call it 'the nomological account'—which can be used to rebut content property epiphenomenalism. Fodor includes, among the laws to which his account applies, laws which are not strict but which contain a *ceteris paribus* clause. This is important with respect to content properties, since it is widely agreed that there are no strict causal laws whose antecedents are content properties. According to Fodor, a property F is causally relevant in a particular causal transaction, c causing e, just in case there is a causal law

---

[6] Kim is aware of this problem. In Kim (1990) he suggests introducing a contextual notion of supervenience so that, although x's having a certain thought doesn't supervene on any causal property that x instantiates, it might supervene on a causal property relative to x's history.

to the effect that Fs cause Gs (*ceteris paribus*), which subsumes *c* and *e*, whose *ceteris paribus* conditions (if there are any) are, in this instance, satisfied.

Fodor and McLaughlin's proposal can be thought of as attempting to resolve our paradox by rejecting step (3) in the paradoxical argument.[7] To rebut property epiphenomenalism, Fodor needs to show that predicates expressing content properties occur in genuine causal laws; that such laws can determine causal relations; and that their doing so is compatible with non-reductive physicalism. We will argue that his account doesn't succeed. First, we will give reasons to think that, to the extent that the nomological account affords a significant causal role to non-basic properties, it compromises physicalism. Second, we will observe that Fodor himself has provided persuasive reasons for thinking that there are no genuine causal laws involving ordinary content properties.

Fodor argues that his account assigns a significant causal role to a property—even to a non-basic property—F which occurs in the antecedent of a genuine causal law, since F can affect the causal powers of events which possess it. That is, it can affect what causal relations *c* enters into. If *c*'s being F is causally relevant to *c*'s causing *e*, then *c*'s being F is, in the circumstances (that is, in circumstances in which the *ceteris paribus* conditions are satisfied), *sufficient* for *c* to cause *e*. Thus it seems that F has the 'power' to determine the causal relations of events which instantiate it.

Fodor's proposal faces the following problem. Suppose that *c* and *e* are subsumed by the *ceteris paribus* law that Fs cause Gs, and that the law's *ceteris paribus* conditions are satisfied. Will there also be a fundamental law of physics which subsumes these events? If Fodor's answer is 'no', then the account is incompatible with reductive physicalism, since there will be causal relations which do not supervene on basic physical properties and laws. If his answer is 'yes', then the question arises of whether it is *c*'s possessing the property F that determines its causal relations or whether its possessing some basic physical property P is the thing in virtue of which it is subsumed under a basic physical causal law that determines its causal relations. That is, is it F or P that gives *c* its causal powers? Presumably Fodor's answer will be: 'both'! That is, *c*'s being P and *c*'s being F, each, have an equal claim to be responsible for *c*'s causing *e*. But we wonder whether physicalism can permit a non-basic property and law to have the same status as basic properties and laws with respect to causation. It appears to us that non-reductive physicalism affords priority to basic physical properties over other properties when it comes to grounding causal relations. Our reason for saying this is that the ability of an F—a non-basic property—to cause a G is itself ultimately grounded in the physical properties of the events. The fact that 'Fs cause Gs' is a causal law is a result of the basic causal laws and of the way in which Fs and Gs supervene on physical facts. So the real locus of causal powers is the physical properties. F, so to speak, gets carried piggyback

---

[7] It appears that this burden will be even more difficult for McLaughlin to discharge, since he is attempting to show that the causal potency of content properties is compatible with Davidson's view that there are not even one-way bridge laws from physical to content properties.

on physical properties, and it is mere appearance that possessing F determines c's causal powers. The basic physical properties and laws determine both the causal relations among events and the non-basic causal laws. It is merely an appearance that the non-basic causal laws determine casual relations among events.

Fodor might grant that c's possessing the non-basic property F determines c to cause e (a G), given the law *Fs cause Gs ceteris paribus*, only in a derived sense, and that this derived sense is still sufficient to supply F with causal potency. But, if F is to have even a derived status in determining causal relations, then it seems, as we suggested in the first section, that there should be a physicalistic account of why the non-basic law holds in this case. If there is not such an account, then we are faced with the causal relation being overdetermined in a way that hardly seems compatible with physicalism. Now Fodor seems to think that there are, in principle, physicalistic explanations of content properties. But if one thinks that there is no such account, then Fodor's strategy doesn't solve the epiphenomenalist paradox.

In a way, the previous discussion concerning Fodor's proposal for characterizing causal potency is moot, since Fodor has argued vigorously and persistently that ordinary content properties, which he calls *broad* content properties, do not, after all, affect causal powers (see especially Fodor, 1987, Ch. 2). By 'broad content' Fodor means externalist content; that is, content which does not supervene on neurophysiology. As we have noted, he argues that due to this failure of supervenience, broad content properties do not individuate states in accordance with causal powers. He goes on to argue that, since genuine sciences always taxonomize with respect to causal powers, intentional psychology must be reformed if it is to be integrated into science. His well-known proposal is to construct 'narrow content' properties for the purposes of individuating intentional states. Narrow content properties supervene on neurophysiological properties; and so, if Oscar satisfies one, so do all his twins.[8]

There is an obvious tension between the view that broad content properties do not affect causal powers and the view that a property affects causal powers if it appears in the antecedent of a causal law. The tension can be removed easily enough by claiming that, despite appearances, there are no broad content causal laws. We suppose that this is Fodor's view. But, if it is, then he has not provided an account of the causal potency of ordinary content properties, since these are broad.

An account in terms of subjunctive conditionals

Supervenient causation is too weak a notion to provide content properties with a sufficiently strong causal role. If, as Fodor hopes, it could be shown that content properties determine causal relations, then they would possess a sufficiently strong causal role; but we have cast doubt on the suggestion that this is compatible with physicalism. We want now to consider another way of understanding 'causal potency.'

---

[8] The precise characterization of narrow content properties and the question of whether or not they are really intentional properties need not concern us here.

When explaining why one thinks that possession of a content property makes a causal difference one is likely to make reference to certain subjunctive conditionals. For example, suppose that Clyde has decided to get some beer and believes that there is beer in the fridge. In these circumstances it would be natural to think that the following (and related) subjunctives are true:

> If Clyde were to think that the beer is in the fridge, then that would cause him to go to the fridge.
> If Clyde were to think that the beer is in the cupboard, then that would cause him to go to the cupboard.

We will call subjunctive conditionals of the form 'if A were to be the case, then that would cause B to be the case' *causal conditionals*. At any time *t* there will be many such causal conditionals which hold of Clyde. We suggest that these and related conditionals lie at the heart of the view that content properties are causally potent. To apply this idea to characterizing causal potency, we will define a relation between event property pairs which, after Terry Horgan,[9] we call 'quasation' (as in *c* qua F causes *e* qua G). As a tentative account of the quasation relation we propose the following:[10]

> $\langle c,F \rangle$ is quasally related to $\langle e,G \rangle$ iff *c* and *e* occur and are respectively F and G and there is some time before the occurrence of *c* at which these two conditionals obtain:
> (1) if *c* were to occur and be F then that would cause an event *e* to be G;
> (2) if *c* were to occur but not be an F then it would not cause an event which is G.

This proposal doubtless requires refinement. For example, we probably want to require that the properties F and G are not analytically or metaphysically connected to each other in order to express the idea that the connection between F and G is a *causal* connection. Some modification to deal with genuine causal overdetermination is also required. There are also problems concerning event individuation and the truth conditions of temporally indexed subjunctive conditionals which need to be addressed.[11] But the idea behind the account is straightforward. In the possible worlds which are most similar to the actual world at time *t* at which property F is instantiated by event *c*, *c* causes an event which is G and in worlds in which c fails to instantiate F it also fails to cause a G.

Let's see how this account fares with respect to the conditions we have laid down on a satisfactory account of causal potency. First, we want to show that content properties can appear in subjunctive conditionals (1) and (2) in the account of quasation above and that their doing so doesn't compromise the principles of non-reductive physicalism

---

[9] Horgan,(1989). Our account of 'quasation' is similar to his, which in turn is similar to the account we suggest in Chapter 11 of this volume.

[10] A slightly different counterfactual proposal is discussed in Chapter 11 of this volume.

[11] Temporally indexed subjunctive conditionals have not received much discussion. The pioneering paper is Gupta and Thomason, 1981.

even if content properties are not reducible to, or explicable in terms of, causal physical properties. To demonstrate this in full, we would need to show that there is no incompatibility between subjunctive conditionals (1) and (2) globally supervening on physical facts (as required by non-reductive physicalism), and the property F possessing the features of content properties. While we cannot do that here, we do want to sketch an account that has has these features.

Suppose that Clyde's thinking that there is beer in the fridge fails to supervene on his neurophysiological properties. This doesn't seem to provide any bar to the counterfactuals which occur in the definition of quasation obtaining. If there is a problem, it is connected with the second counterfactual. The worry is that, if Clyde were not thinking that there is beer in the fridge, then he would still go to the fridge, because he is thinking that there is a twin beer in the fridge. But, if Clyde is anything like us, then, in all the closest worlds in which he is not thinking that there is beer in the fridge, he is thinking that it is in the cupboard—or some such thing. In any case, the world in which he is thinking that there is a twin beer in the fridge is certainly a very distant one.

As far as we can see, the non-existence of an explication of content properties in physicalistic terms doesn't present an obstacle to the causal potency (defined in terms of the quasation relation) either. If there is no physicalistic explanation of content properties, then there will not (generally) be a physicalistic explanation of the counterfactuals in which they occur. But, unlike with genuine causal laws we see no reason why counterfactuals must be physicalistically explicable if they are to be true. It is sufficient that the counterfactuals globally supervene on physical facts; and this is secured by assuming that worlds which are exactly alike with respect to physical facts possess the same similarity relations to other worlds.

The questions remain of whether or not $\langle c, F \rangle$ is quasally related to $\langle e, G \rangle$ and whether the related notion of causal potency are sufficiently strong to rebut epiphenomenalism. First, it should be clear that we are not offering this account of the quasation relation as a way of claiming—as Fodor does—that content properties determine an event's causal powers. As far as we can see, an event's causal powers are completely determined by its basic causal properties. Content properties are not needed for that. However, if $\langle c, F \rangle$ is quasally related to $\langle e, G \rangle$, then there is a perfectly good sense in which c's having F *makes a difference* to c's causal powers. This is the sense captured by the counterfactuals (1) and (2). To say that instantiating F makes a difference to what c causes is to say, in part, that, had c not had F, then it would not have caused a G. Of course, this counterfactual is true in virtue of certain physical facts obtaining. But that doesn't make these conditionals any less true or any less explanatory.

Acknowledgments

We would like to thank Louise Antony, Paul Boghossian, Jerry Fodor, Joe Levine, Brian McLaughlin, and Sigrun Svavarsdottir for discussions of content causation issues and for comments on an earlier version of this chapter.

References

Block, N. (ed.) (1980) *Readings in Philosophy of Psychology*, Vol. 1. Cambridge, MA: Harvard University Press.

Boghossian, P. (1990) Naturalizing Content. In Barry Loewer and Georges Rey (eds): *Meaning in Mind: Fodor and his Critics*. Oxford: Basil Blackwell, pp. 65–86.

Dretske, F. (1981) *Knowledge and the Flow of Information*. Cambridge, MA: MIT Press.

——(1987) *Explaining Behavior*. Cambridge, MA: MIT Press.

Dretske, Fred (1989) Reasons and causes. *Philosophical Perspectives* 3: 1–15.

Fodor, J. (1987) *Psychosemantics*. Cambridge, MA: MIT Press.

——(1989) Making Mind Matter More. *Philosophical Topics* 67/1: 59–79.

——(1990) A Theory of Content Part II. In *A Theory of Content and Other Essays*. Cambridge, MA: MIT Press.

Gupta, A., and R. Thomason (1981) A Theory of Conditionals in the Context of Branching Time. In W. Harper, R. Stalnaker, and G. Pearce (eds), *Ifs*. Dordrecht: D. Reidel, pp. 65–90.

Horgan, T. (1989) Mental Quasation. In J. Tomberlin (ed.), *Philosophical Perspectives* 3: 47–76.

Kim, J. (1983) Supervenience and Supervenient Causation. *The Southern Journal of Philosophy* 22 (Supplement 1983): 45–56.

——(1984) Epiphenomenal and Supervenient Causation. *Midwest Studies in Philosophy* 9: 257–70.

——(1990) Explanatory Exclusion and the Problem of Mental Causation. In Enrique Villanueva (ed.), *Information, Semantics, and Epistemology*. Oxford: Blackwell, pp. 36–56.

Loewer, B. (1987) Information and Intentionality. *Synthese* 70: 287–317.

McLaughlin, B. (1983) Event Supervenience and Supervenient Causation. *The Southern Journal of Philosophy* 22 (Supplement): 71–92.

——(1989) Type Epiphenomenalism, Type Dualism, and the Causal Priority of the Physical. In J. Tomberlin (ed.), *Philosophical Perspectives* 3: pp. 109–135.

Philosophy of Mind and Action Theory, Atrascadero: Ridgeview Publishing Company, pp. 109–135.

13

From Physics to Physicalism

Barry Loewer

Introduction

Hilary Putnam explains that

[t]he appeal of materialism lies precisely in this, in its claim to be *natural* metaphysics within the bounds of science. That a doctrine which promises to gratify our ambition (to know the noumenal) and our caution (not to be unscientific) should have great appeal is hardly something to be wondered at. (Putnam, 1983, p. 210)

Materialism says that all facts, in particular all mental facts, obtain *in virtue of* the spatio-temporal distribution, and properties, of matter. It was, as Putnam says, 'metaphysics within the bounds of science', but only so long as science was thought to say that the world is made out of matter.[1] In this century physicists have learned that there is more in the world than matter and, in any case, matter isn't quite what it seemed to be. For this reason many philosophers who think that metaphysics should be informed by science advocate *physicalism* in place of materialism. Physicalism claims that all facts obtain *in virtue of* the distribution of the fundamental entities and properties—whatever they turn out to be—of completed *fundamental physics*. Later I will discuss a more precise formulation. But not all contemporary philosophers embrace physicalism.

Some—and, though these are a minority, it is not a small or uninfluential one—think that physicalism is rather the metaphysics for an unjustified *scientism*, in other words scientistic metaphysics. Among those who hold that physicalism can be clearly formulated, many think that it characterizes a cold, colorless, unfeeling, and uninteresting world and not the world we live in. In their eyes it is—or it would be, if they thought we had reason to believe it—a doctrine to be feared. In this chapter I will argue that these fears have little foundation. (Perhaps they result from traumatic exposure to other doctrines called 'physicalism'.)

---

[1] Putnam himself is no advocate of materialism or physicalism, but he thinks that metaphysical claims like these involve presuppositions he rejects. As will become clear, while I like the quote, I disagree with his view.

Formulating physicalism

Physicalism is sometimes formulated (for instance by Crane, 1991) as the thesis that all that God had to do in order to create our world was to create its physical facts and laws; the rest followed from these. Fortunately for physicalism's proponents, there are non-theological formulations. The following is due to Frank Jackson (1998):

> (P) Physicalism is true IFF every world that is a minimal physical duplicate of the actual world is a duplicate *simpliciter*. (1999, p. 56)

In (P), a physical duplicate duplicates the laws of physics as well as the physical facts. A minimal physical duplicate duplicates just this and nothing more than what is absolutely (i.e. metaphysically) necessary. The idea behind (P) is that, once one has fixed the physical facts of our world, one has thereby, metaphysically speaking, fixed all the facts. Notice that (P) is itself not metaphysically necessary. There are possible worlds that contain non-physical entities or properties. Minimal physical duplicates of such worlds are not duplicates *simpliciter*.[3]

As Jackson notes, the truth of (P) is necessary to capture the idea that all facts hold in virtue of physical facts. If (P) is false, then there is a world that completely matches the actual world physically, but leaves something out. If there is something over and above the physical that would thus be left out, obviously there are some facts that do not hold in virtue of physical facts. On the other hand, it is not clear that (P) is sufficient for physicalism. The worry is that (P) fails to capture the idea that the fundamental properties and facts are physical and everything else obtains *in virtue of* them. One problem is that (P) does not exclude there being some kind of fact other than physical facts, such that minimally duplicating this kind duplicates the world. It helps to add to (P) the claim that there is no other kind of fact, and I will henceforth understand (P) with this addition. But this still may not be enough. The worry is that (P) may not exclude the possibility that mental and physical properties are distinct but necessarily connected in a way in which neither is more basic than the other. In this case, it doesn't seem correct to say that one kind of property obtains in virtue of the other's obtaining.[4]

---

[2] Lewis (1983) formulates physicalism as the doctrine that 'among worlds where no natural properties alien to our world are instantiated, no two differ without differing physically'. A property is alien to a world *w* if it is not instantiated in *w* and not constructed out of properties instantiated in *w*. On the assumption that fundamental properties of physics are natural properties, Lewis' formulation entails Jackson's; but the converse entailment doesn't hold, at least not without further metaphysical assumptions about natural properties and laws. While both formulations employ possible worlds, neither is committed to Lewis' account of possible worlds.

[3] (P) is contingent but, in a certain sense, non-accidental. If (P) is true and if Q is any proposition (true or false) that is compatible with the laws of physics, then the counterfactual "if Q were true then (P) would be true" is also true. See Witmer (2001) for a discussion of this point.

[4] How might this happen? One way would be if properties were individuated by their nomological connections to other properties (Shoemaker, 1998) so that, e.g. F cannot be instantiated without G's being instantiated. Another way (Russell, 1927) would be if fundamental properties possess both categorical and nomological/causal aspects that are metaphysically inseparable and if physical concepts refer to the

Now I am not sure that this is a real possibility since it may be that, if two *fundamental* properties are really distinct, then they cannot be connected by metaphysical necessity.[5] But, if considerations about the nature of necessity do not rule out this possibility, then we must admit that (P) is not quite sufficient for physicalism. However, it seems to me that, if we had good reason to believe (P), then, unless we also had some reason to believe that, despite (P), mental facts (or some other kind of facts) do not hold in virtue of physical facts, we have good reason to accept physicalism. In any case, even if (P) is not quite physicalism, it is close enough to be an interesting claim. So for the rest of this chapter I will focus on the credibility of (P).

Some philosophers dismiss physicalism as a serious view because they are skeptical about the possibility of characterizing adequately the notion of a *physical* statement. The alleged problem (Crane and Mellor, 1990) is that every way of explicating 'physical' will make (P) either obviously false or trivially true (or not consonant with the idea behind physicalism). If 'physical' is characterized in terms of the language of current physics, then it is likely that (P) is false, since it is likely that the vocabulary of current physics is incomplete.[6] Certainly a complete description of the world in terms of the vocabulary of classical physics is incomplete, and there is reason to suspect that there are additions (if not revisions as radical as the replacement of classical physics by quantum physics) that will be made to physics in the future. On the other hand, if 'physical' in (P) means facts expressible in the language of the complete physical theory of the world (if there is one), then that threatens to make (P) trivial, unless some conditions are placed on what makes a theory 'physical'. If it were to turn out that to account for certain clearly physical events physicists needed to posit fundamental intentional, or phenomenal, properties, then the resulting theory would not be physical.

The most straightforward response to this objection (Papineau, 1993) is to require that fundamental physical predicates (the atomic predicates of the language in which the complete physical description of the world is expressed) are not mental (that is, intentional and phenomenal). Given this, (P) implies that mental truths supervene on non-mental truths. This is non-trivial and not obviously false. Other and stronger versions of physicalism can be formulated by adding further conditions on fundamental physical predicates, for example, no biological predicates, no macroscopic predicates, and so on. Another route is to add to (P) the claim that our world can be fully described

nomological/causal aspect but not to the categorical aspects. Some philosophers inclined to this view think of the categorical aspect as proto-mental, and others as merely non-physical and unknowable. In either case, if the properties referred to in physics possess this kind of categorical aspect, it seems wrong to say that all facts hold in virtue of physical facts even though (P) is satisfied.

[5] According to Lewis' 'recombination principle' (1983), if P and Q are distinct fundamental properties, then there are possible worlds in which one is instantiated but the other is not.

[6] The trouble isn't that current physics may be false (i.e. physicists are mistaken about the laws), but that it may be incomplete. If its vocabulary were sufficiently complete to specify all the fundamental physical facts and laws, then (P) would not be obviously inadequate.

using only spatio-temporal and microscopic predicates (in addition to logic and mathematics); we can call a theory of this sort a 'PC theory'. So (P) can be taken as formulating a number of physicalist doctrines, the strongest one identifying the physical vocabulary as constructed from the predicates of the true PC theory. In what follows I will generally have this strongest version in mind.[7]

> (P) has an important consequence: what Jackson calls 'the entry by entailment thesis'
> A first, but not quite correct, formulation is:
> (ENT) If B is true and $ is the full physical description (including the laws) of the world, then N(If $, then B).

ENT isn't quite correct, since, even if (P) is true, there will be some true statements, say, 'there are no spirits', that are not entailed by $. The problem is that there are worlds at which $ is true but which contain non-physical entities (such as spirits) or at which non-physical alien properties are instantiated. Jackson (1998) recognizes this problem and seeks to handle it by requiring that $ include not just the full physical description of the world but also a statement that says, as he puts it, 'that's all.' But it is not clear how to formulate 'that's all' without going beyond purely physical and mathematical vocabulary.[8] A better approach is to restrict the statements B that appear in ENT to ones that are in an intuitive sense 'non-global' (Witmer, 1997). Global statements are those whose truth-values depend not just on the fundamental facts, but also on a given collection of fundamental facts being *all* the fundamental facts that obtain at a given world. We can make this precise by defining a proposition as global just in case it is true at some world w and by specifying that, for any world w at which it is true, if world $w^\star$ contains w, then the proposition in question is also true at $w^\star$. 'There are no spirits' and 'there are exactly ten conscious beings' are examples of global propositions, while 'there are at least ten conscious beings in region R' and 'John is in pain' are examples of non-global statements.[9] Many of the basic predicates of the special sciences are non-global, in that statements affirming their instantiations are non-global. However, a true generalization of a special science may well be global. If so, then, although it supervenes (and, if it expresses a law, the law supervenes) on the physical facts, it will not be entailed (and its being a law will not be entailed) by the full physical description.[10]

---

[7] This strong version has the apparent disadvantage of defining physicalism so that it entails PC; but since I think that PC is plausibly true and that it provides the premise for the best argument for physicalism, this doesn't strike me as a real disadvantage.

[8] Adding that '$' is the complete description of the world involves semantic notions.

[9] Some global statements, e.g. 'there are no ghosts', have negations that are non-global, while others, e.g. 'the average age of a ghost is a million years', have negations that are also global.

[10] If the reduction of a special science to physics involves the laws of physics and statements of physics implying the laws of the special science, then (P) doesn't imply that the special science laws are reducible to physics.

Jackson thinks that, if physicalism is true, then true instances of ENT are analytic and a priori. His argument for this surprising claim involves a (controversial) generalization of Kripke's account of how certain statements express necessary truths and yet are only knowable a posteriori, for example 'water $= H_2O$'. In fact, at one time, Jackson thought that statements concerning phenomenal consciousness do not follow a priori from the full physical description, and that this showed that physicalism is false.[11] But Jackson's argument involves semantical assumptions that are no part of physicalism and, in any case, it can be demonstrated that the claim has exceptions.[12]

It is interesting to compare Jackson–Lewis physicalism with another influential view in the philosophy of mind, which is considered physicalist by some and anti-physicalist by others. I have in mind Davidson' anomalous monism (henceforth AM). AM consists of three claims:

(1) every event is a physical event;
(2) there are no strict psycho-physical or psychological laws;
(3) psychological predicates supervene on physical predicates.

None of these claims is free from obscurity. Davidson's conception of events is one on which a single event can satisfy many descriptions of many different kinds, such as the pulling of the trigger $=$ the killing of the Archduke $=$ the starting of WWI, and so on; and in particular it can satisfy both mental and physical descriptions. Events are particulars with location in time and space (modulo the same kinds of vagueness possessed by objects). All that Davidson says about the individuation of events is that, if c and $c^\star$ have the same causes and effects, then they are the same event—but this is not very helpful, since causes and effects are themselves events. He also seems to think either that all of an event's properties are essential to it or—more charitably—that talk of whether a particular event could have lacked a certain feature—or failed to satisfy a certain description—is not really sensible. This view contrasts with views of Kim, Lewis, or Yablo, on which events are, or correspond to, instantiations of properties at times (or over durations).

Exactly what Davidson thinks that a law is is also obscure; and even more obscure is his argument for the non-existence of psycho-physical laws. But it is clear that he thinks those arguments to establish that 'there are no tight connections between the mental and the physical' (1970, p 117). In particular, he would certainly reject the view that the physical state of a person, and her environment, and the laws of physics necessitate her mental state.

Davidson's supervenience thesis is also unclear. He has waffled between the claim that in the actual world (or in each world) no two individuals (objects or events) can differ

---

[11] Jackson, (1998). Chalmers (1996) develops this argument in great detail.
[12] For general objections to Jackson's claim, see Yablo, 2000. For an argument that shows that the claim must have at least some exceptions, see Balog, 2001. Balog shows that statements about phenomenal consciousness must be exceptions to the claim.

mentally without differing physically (weak supervenience) and the much stronger claim that for any loss of physically possible worlds individuals from those worlds can differ mentally without differing physically (strong supervenience). It has been argued that the latter claim is sufficient for there being psycho-physical laws. Whether or not that is right depends on exactly what laws are, but it is certainly sufficient for the physical properties of an individual A and for the laws of physics to necessitate A's mental properties. So this reading would seem to conflict with the 'anomalous' part of AM. Weak supervenience is very weak and very trivial if spatio-temporal locations are included among physical predicates. If they are not included, then weak supervenience is stronger, but it is very difficult to motivate why it should hold.

In any case, if AM is understood as denying the existence of tight connections of the sort entailed by Jackson's physicalism, then it is weaker than it and not physicalist at all, since we say that Jackson's formulation is at least necessary for physicalism. Further, the event identity thesis is neither necessary nor sufficient for physicalism. Event identity isn't sufficient since it may be that there are mental properties whose instantiations are not physically necessitated. It is not necessary since it may be that the modal properties of mental and physical events differ, even though the existence and the modal properties of mental events are physically necessitated. Finally, if the supervenience thesis is understood as strong supervenience, then (if understood in the sense that all instantiated properties strongly supervene on physical properties) the thesis is stronger than Jackson's physicalism—which is unreasonably strong—since it means that, as a matter of metaphysically necessity, there cannot be a world whose sole inhabitants are souls!

Fear of physicalism

I now want to consider some alleged consequences of physicalism:

> Physics takes precedence over all the other sciences and other ways of obtaining knowledge.
> All sciences (and all other truths) are reducible to physics.
> The only genuine properties, events, and individuals are those of fundamental physics.
> The only genuine laws are laws of physics. There are no special science laws.
> The only genuine causation is causation by physical events; in particular, there is no mental or other, higher-level causation.
> Eliminativism: intentionality, consciousness, rationality, freedom, and norms do not exist.

Sometimes these alleged consequences are taken to be reasons to disbelieve physicalism, and sometimes unpleasant truths that physicalists must learn to live with. I agree that each of the above claims is indeed unpalatable; but none is a consequence of (P). Decisively establishing that this is so is beyond the scope of this chapter, but a brief discussion may lend some credibility to this view.

1. It is important to keep in mind that (P) is not an epistemological or methodological doctrine. (P) doesn't imply that physics is epistemologically or methodologically more basic than other sciences. If the claims of current physics should conflict with the claims of one of the special sciences, it may very well be that the latter is better confirmed than the former. Nor does the truth of (P) favour allocating federal funds to the proposed super-collider over biological research, or to support the arts. (P) doesn't imply that the reductionist methodology of analyzing complex systems into simpler parts, which is so successful in physics, is the appropriate method for the enormously more complex systems treated in biology. Nor does (P) imply that scientific methodology is the only, or the best, way to acquire knowledge about every topic. It is compatible with (P) that hermeneutics, or *verstehen*, is a better method than experimental psychology when it comes to knowing the minds of our fellow human beings. (P) does require that, if that is so, then this fact, like all others, is implied by the totality of physical facts.

2. A reduction of a special science (say, biology), or of a particular theory of a special science, to fundamental physics involves systematically locating physical truths that entail the truths (including the laws) of the special science. (P), of course, does not imply that any special science can be reduced to current physics. More interestingly, it doesn't imply that any of the special sciences can be reduced to completed physics. Since laws are expressed by global statements, (P) doesn't imply that physical truths imply special science laws. And, although (P) does imply that non-global truths of the special sciences are metaphysically entailed by the truths of fundamental physics, it does not imply that these entailments are systematic, or that we can ever locate and know them. Similarly, (P) does not imply that truths that do not belong to any science can be reduced to physics, even though they are implied by statements of physics. It may be, as some have suggested, that limitations in the kinds of concepts that we are capable of entertaining, or in the nature of the concepts of the special sciences, or in the complexity of the entailments, preclude us from ever knowing the implications that (P) requires.[13]

3. (P) does imply that the only *fundamental* properties, events, and individuals are those of fundamental physics. But (P) doesn't exclude non-fundamental entities belonging to these various categories. Exactly what non-fundamental properties (P) allows for depends on the nature of these properties. Let me begin with what is sometimes called an 'abundant' conception of properties. According to the abundant conception, every predicate (or concept) that can be used to make a statement (or is a constituent of a thought) with truth conditions expresses a property, although there may also be properties that are not expressed by the predicates of any language.[14]

---

[13] McGinn (1999) claims that limitations in our concepts may prevent us from ever solving certain philosophical problems, e.g. how phenomenal consciousness arises out of physical phenomena; Loar (1997) and Balog (2001) suggest that the nature of the phenomenal concepts themselves prevents our finding any identification or realization of phenomenal consciousness by physical properties satisfying.

[14] For Lewis (1983) an abundant monadic property is any set of possible entities.

Further, two predicates express the same property iff it is metaphysically necessary that they are co-extensive. Although the abundant conception is profligate with respect to the existence of properties, it is compatible with properties being non-linguistic and (largely) mind-independent, and with distinct predicates expressing the same property.[15] It is clear that, on the abundant conception, (P) is compatible with the existence not only of the properties of fundamental physics, but also with many other kinds of properties. On the abundant conception, 'is a storm', 'is green', 'is five miles from the Eiffel tower', 'is grue', 'feels painful', 'is a thinking about Vienna', all express properties. It may be that, given (P), some of these properties are not instantiable, but that would have to be shown for the specific property.

Contrasting with the abundant conception are the so-called 'sparse' conceptions of properties, according to which only certain predicates express genuine properties. Lewis calls these 'perfectly natural properties' (Lewis, 1983; see also Armstrong, 1978).[16] Perfectly natural properties satisfy certain conditions, for instance they involve real similarities, they figure in laws and scientific taxonomies, and they are causally relevant. Such added conditions presumably disqualify 'grueness' (and perhaps some of the other examples mentioned above) from being a perfectly natural property. So, whether or not the instantiation of perfectly natural properties that are not expressible in the language of physics (in particular psychological properties) is compatible with (P) depends on whether the existence of special sciences (in particular psychology) and higher-level causation is compatible with P; issues to which I will turn below.

Suppose for now that (P) doesn't exclude the existence and instantiation of perfectly natural properties other than those of fundamental physics. How might these properties and their instantiations relate to physical properties? If (P) is true, then these instantiations (for the non-global properties) are necessitated by physical facts and laws. But in some cases more can be said. It is widely and plausibly held that some non-fundamental properties are *realized* by fundamental physical properties. There are various views concerning exactly what is involved in one property instantiation realizing another, but they have it in common that, if an instance of P realizes an instance of F, then the P instantiation metaphysically necessitates the F instantiation; in other words any possible world that contains the first also contains the second.[17] For example, the property of being a storm may be realized by various dynamic configurations of physical properties, but it is not identical to any specific one, or even to their disjunction. For it seems plausible to suppose that storms occur in possible worlds that

---

[15] It is important to distinguish between concepts and properties. Concepts are the meanings of predicates, while properties are the references of predicates. A thinker may understand two distinct predicates without realizing that they express the same property, e.g. 'is water' and 'is H_2O'.

[16] Universals can be construed as special abundant properties, or as another kind of fundamental entity, which is correlated with an abundant property. Lewis (1983) points out that the work that is done by universals can also be done by tropes.

[17] Shoemaker (2001) suggests that a property P realizes F when the set of F's causal powers is a subset of the set of Ps causal powers.

FROM PHYSICS TO PHYSICALISM 203

contain none of our fundamental physical properties (these storms will be realized by alien fundamental properties).

Suppose that events are, or are correlated with, the instantiations (by individuals and times) of certain event-constituting properties (Kim and Brandt, 1967). I see no reason why non-fundamental properties cannot be event constituting. For example, the event of a storm striking the coast is composed of various physical events, though it is somewhat vague around the boundaries; but is not identical to any of them. This is clear, since the storm's modal properties differ from the modal properties of the events that compose it. Of course, these counterfactual truths must, if (P) is true, supervene on the totality of truths of fundamental physics.[18]

Similar remarks apply to particulars, for instance to particular rocks, trees, and people. If (P) is true, then these 'higher-level' particulars are constituted by the instantiations of physical properties, elementary particles possessing certain states; but they are not identical to them. Of course, exactly what fundamental physical entities constitute a given particular, say, my cat, is a vague matter. The important point is that (P) is compatible with the existence of my cat even though the cat is not identical to any fundamental physical particles (or mereological sum of them). The reason why my cat is not identical to any sum of fundamental physical entities is that there are counterfactuals which are true of my cat but are not true of the mereological sum (or any number of more precise versions of the vague sum) that constitute it. Of course, as in the case of events, any such counterfactuals must supervene on the totality of physical facts.

4. Of course, (P) is not compatible with the existence of *fundamental* laws that are not laws of physics. Whether or not (P) is compatible with there being non-fundamental laws depends on what exactly laws are; and that is controversial.[19] Jaegwon Kim (1993) has argued that physicalism excludes there being special science laws. More precisely, he argues that, if F and G are non-physical predicates each of which is multiply realized, then the generalization 'Fs are followed by (or cause) Gs' cannot express a law. His argument is predicated on the idea that statements of law must involve predicates that are projectable and, if physicalism is true, then only predicates of physics are projectable. 'A predicates G' is projectable with respect to F (the generalization 'Fs are followed by Gs' is projectable) if Fs which are Gs confirm that Fs are followed by Gs.[20] Since he holds that a generalization expresses a law only if its predicates are

---

[18] The same holds for other accounts of events, e.g. Lewis (1986) Yablo (1992) and Davidson's (1969) rather different account. An interesting point about Davidson' account is that his event monism, all events are physical events, is neither necessary nor sufficient for (P).

[19] It is plausible that special science laws or law statements include a *ceteris paribus* qualification. There is no fully satisfactory account of exactly what this qualification comes to; but that shouldn't prevent us from recognizing that there are plenty of examples of special science statements that are thought to express laws.

[20] Following Goodman (1979), a hypothesis of the form 'Fs are followed by Gs' is projectable (G is projectable with respect to F) if and only if positive instances of the hypothesis confirm it (raise its credence and the credence of unobserved positive instances).

projectable, he concludes that (P) is incompatible with the existence of laws expressed by generalizations composed of non-physical predicates. Kim's main argument involves an example. He asks us to consider the property of being jade, which is realized by, or equivalent to, the disjunction jadeite or nephrite. Kim observes that, if our evidence consists just of samples of Jadeite and we are interested in a generalization like 'jade melts at such and such a temperature', then it would be foolish to generalize just from this sample. Of course this is correct, since we know that different molecular structures typically give rise to different melting temperatures. But this example doesn't generalize to other predicates. In the first place, whether or not a generalization is lawlike depends on the predicates both in the antecedent and in the consequent. There are some generalizations with 'jade' in the antecedent that are confirmable by their instances; for example 'jade is called "jade" by English speakers'. In fact, one can see that (P) is compatible with one's assigning a probability function over certain generalizations expressed in non-physical vocabulary, say, the vocabulary of psychology or biology, which allows for their confirmation. That is, predicates of the higher-level sciences may be projectable even though they express multi-realizable properties. Indeed a prohibition against such probability functions would be pig-headed, since we have reason to believe that generalizations that we can express in non-physics vocabulary are sometimes true, or approximately true. We can see that something must be wrong with Kim's argument from the fact that there are well-confirmed generalizations of the special sciences that involve properties that are multi-realized by fundamental physical properties.[21] Perhaps the best and most obvious examples are provided by statistical mechanics—for example, the probability of the entropy of an isolated body decreasing is negligible.

As far as I can see, (P) does not preclude there being true generalizations couched in non-physical vocabulary (whose predicates do not refer to properties of fundamental physics) that are confirmable by their instances, support counterfactuals, and are entailed by highly informative and simple theories, that is, have all the usual marks of laws. If this makes them laws (although non-fundamental laws), then (P) does not exclude special science laws.

5. The issue of whether (P) allows for causation by non-physical events and non-physical properties, in particular by mental events and properties, is a vexed matter. Part of the difficulty is that there are no fully acceptable accounts of causation by either events or properties, even for fundamental events and properties. But it seems to me that any adequate account of causation must allow for causation by non-fundamental entities. For example, the storm's striking the coast caused flooding. The storm is a cause and the property of being a storm is causally connected to the flooding. Of course, such causation is not fundamental, but will supervene on physical facts and laws. As far as I can see, any account that allows non-fundamental physical events and

[21] More extensive responses to Kim's argument are in Block (1997), Fodor (1998) and Antony and Levine (1997).

properties to be causes will do the same for at least some kinds of mental events and properties. I won't argue for that here, but I do want to sketch an account of causation which, although not fully adequate, has this consequence. The account is this: M*s* causes P*s*★ if M*s* does not metaphysically entail P*s*★ and at a time *t* immediately prior to M*s* the following pair of counterfactuals are true:

M*s* >*t* P*s*★;
-M*s* >*t* -P*s*★.

As far as I can see, (P) is compatible with the pair of counterfactuals that, on this account, ground claims such as that the storm's striking the coast caused the flooding. And, additionally, (P) is also compatible with the pair of counterfactuals that ground the causal efficacy of the mental—for example, Fred's wanting a beer causing Fred's body to be in the kitchen. On this account, there may be many events or property instantiations, even ones that occur simultaneously, at different levels that are causally connected to a given event. For example, a neurological event that instantiates Fred's urge may also be causally connected to his body's location.

It has been objected (Kim, 1998) that counterfactual accounts of causation are inadequate, especially in the context of mental causation, since the truth of counter-factuals like the above pair is compatible with epiphenomenalism. That is, it is compatible with there being mental properties that are distinct from the physical properties to which they are connected by so-called 'bridge' laws. If there are no causal relations between these mental properties (or events) and physical properties, then the mental properties are epiphenomenal. This view is non-physicalist, since a minimal physical duplicate of a world at which it obtains does not duplicate the mental property instantiations or the bridge laws. Kim's claim is that, in the epiphenomenalist world, mental to physical counterfactuals may still obtain. But, as we will see in the next section, there is reason to doubt this. In any case, even if it is true, it doesn't show that the counterfactuals are not sufficient for causation in a physicalist world. Indeed, if they or some similar account were not correct, then there could be two PC worlds exactly alike in their physical laws and instantiations of physical and mental properties, but only in one would there be *genuinely causal* connections between mental properties and physical properties. These causal connections would, of course, be themselves non-physical, since they don't supervene on the instantiations of physical properties and laws. But what are these causal connections doing? They are truly 'epiphenome-nal': unneeded and unknowable. The conclusion is that the counterfactual surrogate for causation (or some similar account) is really all we need to account for mental causation as we know it.

Now, of course, Kim would reject the above reasoning, since what he really wants is for mental properties to make a causal difference over and above causation by physical properties. But it is obvious from the start that this is something a physicalist cannot allow. It is enough for our bodily movements to be counterfactually sensitive to our mental states. That the counterfactuals themselves are true in virtue of more fundamental physical facts takes nothing away from that. And one can take comfort in the fact that mental properties are in the same boat as other non-fundamental properties.

The claim that mental causation is compatible with (P) is highly abstract and, of course, doesn't imply that mental causation is implemented by physical processes. Advocates of (P) would like to be in a position to show that certain physical mechanisms do implement mental causation—for example, to explain the causal processes involved when a driver notices a deer running across the road and then applies the brakes. Current psychological research is some distance from coming out with an account, but optimistic cognitive scientists think that the so-called 'computational–representational theory of mind' is on the right track. If intentional causation can be shown to be implemented by computational processes, then we are on the way towards showing that it is physically implemented, since we know that computational processes can be physically implemented—for example, by computers (Rey, 1997).

6. If P really did exclude the instantiation of mental properties (consciousness, intentionality, rationality), then it would have to be rejected, since mental phenomena certainly exist. It is difficult to see how elements of physical reality can add up to intentionality and consciousness. There have been various attempts to show that they do. These 'naturalization' projects have not met with much success.[22] But, as I mentioned previously, (P) doesn't require that we can ever see how physical statements metaphysically entail mental statements, just that the entailments hold. On the other hand, attempts to show that (P) excludes mental facts are, if anything, even less convincing. Here I only want to mention briefly two such attempts. One involves so-called 'conceivability arguments', which claim that it is *conceivable* that the physical facts are what they are and that there is no consciousness, or intentionality, at all.[23] These arguments assume (and sometimes argue for) the claim that, at least in the relevant situations, conceivability implies possibility. However, it can be shown that, at most, conceivability is defeasible evidence for possibility (Balog, 1999). The second line of argument identifies some feature of mental properties, say, alleged normativity, and then claims that it is not the case that physical facts that fail to exhibit this feature can metaphysically entail facts that do. The claim is that two kinds of facts are just too different from each other.[24] While it is not difficult to find philosophers expressing sympathy for this line of thought, it is well nigh impossible to find an argument for this

---

[22] For a survey of some recent attempts to naturalize intentionality, see Loewer (1997).

[23] Conceivability arguments directed at showing that phenomenal consciousness fails to supervene on physical facts go back to Descartes and have recently been revived by Kripke (1980), Jackson (1998), Chalmers (1996), among others. They have, in my view been decisively refuted by Balog (1999) and (2001), Loar (1997), and Hill and McLaughlin (1999).

[24] Davidson's (1969) argument against the existence of strict psycho-physical laws, and more generally against there being any 'close connection' between physical and intentional concepts, seems to be of this sort; but it is very difficult to know this, since it is very difficult to say exactly what the argument is.

claim that is articulated well enough to evaluate. In any case, as we will see, given CP, if any of these arguments were sound, we would be saddled with the conclusion that mental events (or properties) are epiphenomenal, or overdeterminants of physical effects. These consequences seem to me to be so implausible as to cast doubt on the soundness of any argument that claims to show that (P) excludes mentality.[25]

If the previous discussion is on track, then reasonable anti-reductionists, humanists, opponents of scientism, and so on have nothing to fear from (P). But of course the question remains whether they have any reason to believe (P).

From CP to P

What would the world be like if (P) were false? First, there would be certain non-global contingent facts, not metaphysically necessitated by the full description of the physical state. But how would these non-physical facts, and more specifically the properties and entities that constitute them, be related to physical properties and entities? One possibility is that both physical and non-physical properties are instantiated, but there are no laws or metaphysical connections linking them.[26] This view makes it quite mysterious why it is that mental properties are found to be associated with only some physical properties; why, for example, rocks or gasses don't have thoughts. Since there are no laws or metaphysical links that connect these properties, such correlations are merely coincidental. Most opponents of physicalism reject this picture and hold instead that mental (and perhaps certain other properties) are emergent, in a robust sense which involves the existence of *emergent* laws linking them to physical properties.[27] These laws are thought of as vertical, since they link the physical state at *t* with the mental state at *t*. Emergentist views sometimes posit horizontal laws linking instantiations of physical and mental properties with each other at different times. Such laws might ground causal relations between mental and physical events (where events are property instantiations at times). If physics is incomplete, then such mental–physical laws might also be required to give full accounts of physical events (or of their chances). The important point about these emergentist laws is that they are neither among the fundamental laws of physics nor entailed by them and by the physical facts. If our world contains them, then God, when he made the world, had to make them in addition to the physical facts and laws.

---

[25] The arguments that are intended to show that (P) excludes mental facts might actually better be understood as showing, if anything at all, that we cannot see how physical statements metaphysically entail mental statements.

[26] This view is sometimes attributed to Davidson (1969), who on the one hand holds that there are no psycho-physical laws, on the other hand accepts that the mental supervenes weakly but not strongly (as required by (P)) on the physical.

[27] The emergentist laws may be 'deterministic' in that the physical state completely determines the co-temporal mental state or 'indeterministic' in that the physical state only partially determines the mental state or determines the chances of various mental states.

Now I don't think there is much reason to believe that there are emergentist laws, either vertical or horizontal. But not having a reason to disbelieve is not a reason for belief, so I now want to examine an argument for (P). I know of two ways of arguing for (P). One is by finding reductions of particular, higher-level facts (properties, events, and so on) to lower-level, and ultimately physical, ones. While there are some notable reductions (or partial reductions), such as of thermodynamics to statistical mechanics, claims of reduction are usually accompanied by much hand-waving. And, while each successful reduction provides some reason in favor of (P), each failed reduction provides some reason against it. So this piecemeal approach is far from conclusive. A very different line of argument seeks to establish (P) all at once, on the basis of very general considerations about laws and causation. The line of argument I have in mind has been formulated in a number of different ways (see for example McGinn, 1982; Peacocke, 1979; Papineau, 1990, 1993, and 1995 and Loewer, 1995). The version I will discuss proceeds in two steps. The first step argues that any property (or event) whose instantiation at t (or in time interval d) is causally relevant to a physical event at least nomologically supervenes on the physical state of the world at t. The second step argues that the laws characterizing nomological supervenience are no more than the fundamental laws of physics. If we further assume that every property instantiation has a physical effect, and that all non-global facts are determined by some collection of property instantiations, then we get (P). I will assume that the fundamental laws of physics are deterministic. It is not difficult, though a bit messier, to run the argument if the fundamental laws are indeterministic. The main premise of the argument is the deterministic completeness of physics (PCD):

> (PCD) For any distinct times t and t', the physical state $S(t)$ and the fundamental physical laws entail the physical state $S(t')$.

I will say that a property instantiation $M(t)$ is physically detectable by $P^\star(t')$ iff at times immediately prior to t the counterfactuals $M(t) \rangle P^\star(t')$ and $-M(t) \rangle -P^\star(t')$ are true. If $M(t)$ is physically detectable by $P^\star(t')$ and $M(t)$ occurs, then $M(t)$ causes or is actually relevant to $P^\star(t')$ in accord with the account given earlier. Now, suppose that PCD is true so that at times s prior to t the physical state $P(s)$, and the laws entail that $P^\star(t')$ (or $P^\star t'$)). Then, whether or not $M(t)$ occurs $P^\star(t')$ (or $-P^\star(t')$) will occur. So, if $M(t)$ is not connected by law to $P(t)$, it will be physically undetectable. So every property instantiation that is physically detectable at the very least nomologically supervenes on the complete physical state. This is not yet (P). There are two problems:

(1) there may be some property instantiations that are not physically detectable; and
(2) even for physically detectable properties, the most we have shown is that they *nomologically* supervene on the physical state.[28]

---

[28] Witmer (1998), commenting on earlier versions of this argument, made this point about Papineau (1993) and Loewer (1995).

But among the laws there may be laws that are not entailed by the full physical description of the world, including physical laws. As for the first problem, it seems to me plausible that all property instances are either physically detectable or supervene on properties that are physically detectable. But, even supposing this is not so, the following principle still strikes me as quite reasonable:

(U) If some instances of M nomologically supervene on the physical state (or some instances of M supervene on property instantiations that are each physically detectable), then, unless there is some positive reason to think that other instances don't supervene on the physical state, we should suppose that all instances of M do so supervene.

I know of no reason to think of any instances of a non-global property that fails to have instances that are physically detectable (or fail to supervene on property instances that are physically detectable). It follows that it is reasonable to suppose that all property instances nomologically supervene on the full physical state.

Nomological supervenience is compatible both with (P) and with the emergentist account sketched earlier, on which M properties are linked by special 'vertical' laws with physical properties (and perhaps also by horizontal laws with physical and mental properties). To establish (P), then, I need to show that the emergentist account is false. I know of no way of conclusively demonstrating this. However, there are some considerations that make the emergentist picture quite unattractive when combined with (PCD). To begin with, notice that it would be bizarre to suppose that a property like being a rock, or a cloud, is linked by some special vertical law (over and above the laws of physics) to the physical state. If that were so, there would be a possible world physically identical to the actual world; but where in the actual world there is a rock at a certain location, in the other world there would be a cloud. It is only for mental properties that the proposal of special vertical laws has any credibility. But even here the result is peculiar. It would entail not merely that zombies are possible (physical, but not mental duplicates of people), but either that mental properties are completely epiphenomenal or, if they are linked by causal law to physical (or other mental) properties, that the instantiations of these properties are pervasively causally overdetermined. The world may be like that, but I think that simplicity considerations suggest that we don't believe it is pending persuasive arguments against (P).

Furthermore, it is arguable that, if PCD is true, then, on the emergentist account, even with horizontal mental–physical laws, physical events do not counterfactually depend on mental events. So, for example, it will not be the case that the location of my body near the refrigerator will depend on my having desired a beer. Here is why. Suppose that I form the desire to have a beer at t, call this 'M(t)', and that there is a physical state P(t) that is linked by an emergentist law to M(t); in other words 'P -> M'. Suppose now that P(t) leads by the complete deterministic physical laws to P*(t'). Then I claim that -M(t)>-P*(t') is false. The argument assumes Lewis' (1983) account of counterfactuals and, specifically, Lewis' account of world similarity. (It doesn't presuppose Lewis account of possible worlds as concrete entities.) To evaluate the counterfactual, we ask what world (or kind of world) is most similar to the actual world (where the laws of physics are deterministic and there are emergentist vertical laws connecting physical with mental properties). For Lewis, world similarity is evaluated according to two factors: (1) the size of the region in which laws of the actual world are violated; and (2) the size of the departure of perfect match in matters of particular fact. There are two kinds of worlds to consider (where 'L' is the conjunction of the complete deterministic laws of physics and 'V' is the emergentist law linking P with M):

W1: $-M(t)$ & $P(t)$ & L & $-V$
W2: $-M(t)$ & $-P(t)$ & $-L$ & V

If W1 is more similar than W2 to the actual world, then the counterfactual will come out false, since in it $P^\star(t')$ will be true. But if W2 is more similar than W1, then the counterfactual might be true, depending on what physical state occurs at t. At first it may appear that there is a tie in similarity, since both worlds involve the violation of a law, i.e. W1 violates an emergentist law, and W2 a dynamical law of physics. But, on Lewis' account of evaluating similarity, that is not so. The reason is that W2 will differ enormously in matters of particular fact from the actual world, whereas W1 will differ only to the extent required to make $-M(t)$ true. This being so on the emergentist accounts, and assuming that the fundamental laws of physics are deterministic, it follows that physical events don't counterfactually depend on emergent events. Whether or not I want the beer would make no difference to the location of my body. But, of course, it does. And (P) is no impediment to it being so. Under (P), worlds of type W1 are metaphysically impossible, since the physical state P and the physical laws are metaphysically sufficient for $M(t)$. This being so, the counterfactual assumption $-M(t)$ requires some violation of the physical laws. If the most similar world in which $-M(t)$ is one at which $-Q(t)$, such that $Q(t)$ leads by law to $P^\star(t)$, then indeed the counterfactual will come out true. And that scenario is completely compatible with (P), depending on the exact details of the physical realizations of M. So I conclude that (perhaps somewhat surprisingly), if we want the physical to depend on the mental and if we think that the fundamental laws of physics are complete and deterministic, then we should also accept (P).

The credibility of PC

I will conclude with a very brief discussion of the credibility of PC. I think it is fair to say that no one thinks that current physics is complete and that there is not a consensus among physicists or philosophers on whether PC is true. Some physicists, for example Steven Weinberg (1992), think that we are fairly close to a unified physics that would validate PC. But there is a rather contrary view, according to which our world is a much sloppier place than the world of Weinberg's dreams. Nancy Cartwright (1999) is a vigorous advocate of this viewpoint. She has been arguing that the fundamental laws of physics work, to the extent they do, only under very special, contrived conditions found in laboratories. If I understand her correctly the view is that the fundamental laws of physics are woefully incomplete, and what laws there are are best understood as containing *ceteris paribus* qualifications. More fundamental than laws are capacities, for instance the capacity of a charged particle to produce an electromagnetic field. There are many capacities associated with entities and properties at different levels, which, together, determine the tendencies for various courses of events to occur. She suggests that lawful regularities are more or less artifacts of laboratory situations, in which interactions are shielded from the various capacities that normally affect them. And she also thinks that emergent *ceteris paribus* laws may appear at higher levels of description than those of fundamental physics. This is an interesting view, and it would definitely be instructive to think through what a world like that would be like, in contrast with a world at which physics is complete. Cartwright's main argument for her view is in the form of a challenge to proponents of PC. Her example is dropping a dollar bill from St Mark's Tower. Proponents of PC think that the fundamental laws of physics, and to a good approximation Newton's and Maxwell's laws, govern the trajectory of the dollar bill as it floats to the ground, although the use of those laws to predict the trajectory is not possible due to the complexity of the interactions between the dollar bill and air molecules. But Cartwright thinks that this is mere dogma. She seems to suggest that, although 'F = ma' may hold in specialized laboratory conditions, it may be false in this case (presumably some of the changes in motion of the dollar bill are not due to forces).

Cartwright's arguments against PC strike me as very weak. They are merely *sceptical* arguments. She doesn't produce a single case in which 'F = ma' (and other examples of fundamental laws) fails, but rather she claims that extrapolating it beyond controlled laboratory conditions is unwarranted. One would expect that, if a putative fundamental law fails outside the laboratory, it would be possible to find evidence for this fact. A Nobel Prize would be in the offing for the discoverer. Even though the fundamental laws (or the equations believed to approximate them) cannot be used to give detailed explanations in complex situations, they can be used to give approximate predictions, or to be used in connection with statistical models to make statistical predictions, and these are born out. Second, unless we have some specific reason to think that moving outside of the laboratory, or increasing complexity, leads to the failure of these laws, then ordinary principles of scientific inference counsel that we should (counter-evidence pending) suppose that they hold generally. So I think that, while the completeness of physics is a contingent claim, there is now reason to believe that it may well be true and scant reason to think it false.

Conclusion

The journey from physics to physicalism is not an entirely smooth one. As we have seen, the starting place, PC, is plausible, but not obviously true. Given PC, the argument for (P) is fairly straightforward, although it requires that one resist epiphenomenalism and nomological overdetermination. Once we have arrived at (P), we still have not quite reached physicalism, since we saw that there may be ways in which (P) could be true, but still not all facts obtain *in virtue of* physical facts. (P) is thus *almost* as credible as PC. I think it's fair to say that it is more credible than its denial. And, since PC is a scientific claim, physicalism well deserves Putnam's title of 'scientific metaphysics'.

References

Antony, L and Levine, J. (1997) Reduction With Autonomy. *Philosophical Perspectives*: 83–105.

Armstrong, D. (1978) Naturalism, Materialism and First Philosophy. *Philosophia* 8: 2–3.

Balog, K. (1999) Conceivability, Possibility, and the Mind-Body Problem. *Philosophical Review* 108/4: 497–528.

Balog, K. (2001) Commentary on Frank Jackson's From Metaphysics to Ethics. *Philosophy and Phenomenological Research* 62/3: 645–652.

Block, N. (1997) Anti-Reductionism Slaps Back. *Philosophical Perspectives* 11: 107–132.

Cartwright, N (1999) *The Dappled World: A Study of the Boundaries of Science*. Cambridge: Cambridge University Press.

Chalmers, D. (1996) *The Conscious Mind*. Oxford: Oxford University Press.

Crane, T. (1991). All God has to Do. *Analysis* 51 (October): 235–244.

Crane, T. and D. H. Mellor (1990) There is No Question of Physicalism. *Mind* 99/394: 185–206.

Davidson, D. (1963) Actions, Reasons and Causes. *Journal of Philosophy* 60: 685–700; reprinted in Davidson 2001a.

Davidson, D (1969) The Individuation of Events. In Nicholas Rescher (ed.), *Essays in Honor of Carl G. Hempel*. Dordrecht: D. Reidel; reprinted in Davidson 2001a.

Davidson, D. (1970) Mental Events. In Lawrence Foster and J. W. Swanson (eds), *Experience and Theory*. London: Duckworth; reprinted in Davidson 2001a.

Fodor, J. (1998) Special Sciences: Still Autonomous after All These Years. *Philosphical Perspecitves* 11: 149–163.

Goodman, N (1978) *Ways of Worldmaking*. Indianapolis: Hackett.

Hill, C. and B. McLaughlin (1999) There are Fewer Things in Reality than Dreamt of in Chalmers' Philosophy. *Philosophy and Phenomenological Research* 59/2: 445–454.

Jackson, F. (1998) *From Metaphysics to Ethics: A Defense of Conceptual Analysis*. Oxford: Oxford University Press.

Kim, J. (1993) *Supervenience and Mind*. Cambridge: Cambridge University Press.

Kim, J. and Brandt R. (1967). The Logic of the Identity Theory. *Journal of Philosophy* 66: 515–537.

Kripke, S. (1980) *Naming and Necessity*. Cambridge, MA: Harvard University Press.

Lewis, D. (1983) New Work for a Theory of Universals, *Australasian Journal of Philosophy* 61: 343–377.

Lewis, D. (1986) Events. In *Philosophical Papers*, Vol. 2. Oxford: Oxford University Press, pp. 241–269.

Loar, B. (1997) Phenomenal States. In N. Block, O. Flanagan, G. Guzeldier (eds), *The Nature of Consciousness*. Cambridge, MA: MIT Press, pp. 597–612.

Loewer, B. (1995) An Argument for Strong Supervenience. In E. Savellos (ed.), *Supervenience: New Essays*. Cambridge: Needham Heights, pp. 218–225.

Loewer, B. (1997) Freedom from Physics: Quantum Mechanics and Free Will. *Philosophical Topics* 24: 92–113.

McGinn, (1982) *The Character of Mind*. Oxford: Oxford University Press.

McGinn (1999) *The Mysterious Flame: Conscious Minds in a Material World*. New York: Basic Books.

Papineau, D. (1989) Why Supervenience? *Analysis* 49/2: 66–71.

Papineau, D. (1993) *Philosophical Naturalism*. Oxford: Blackwell Publishers.

Peacocke, C. (1979) *Holistic Explanation*. Oxford: Clarendon Press.

Putnam, H. (1983) *Realism and Reason. Philosophical Papers*, Vol. 3. Cambridge: Cambridge University Press.

Russell, B. (1927) *The Analysis of Matter*. London: Kegan Paul.

Shoemaker, S (1998) Causal and Metaphysical Necessity. *Pacific Philosophical Quarterly* 79/1: 59–77.

Shoemaker, S. (2001) Realization and Mental Causation. In C. Gillet and B. Loewer (eds), *Physicalism and Its Discontents*. Cambridge: Cambridge University Press, pp. 74–99.

Weinberg, S. (1992) *Dreams of a Final Theory: The Search for the Fundamental Laws of Nature*. New York: Pantheon Books.

Witmer, D. G. (1998) What is Wrong with the Manifestability Argument for Supervenience? *Australasian Journal of Philosophy* 76/1: 84–89.

Yablo, S. (1992) Mental Causation. *Philosophical Review* 101: 245–280.

14

Mental Causation, or Something Near Enough

Barry Loewer

Descartes claimed that he was able to clearly and distinctly conceive of mind and body apart from each other and concluded that they are distinct substances meeting only in the pineal gland. Ever since then the problem of mental causation has been at the centre of philosophy of mind. Few of Descartes' contemporaries found his glandular proposal for how events involving a non-spatial immaterial substance causally interact with events involving an extended material substance persuasive. Subsequent advances in the physical and biological sciences have convinced many of us that not only is it incomprehensible how a non-material mind can move a material body but that it isn't needed to do so.[1] These and other problems ultimately killed off Cartesian substance dualism. Nevertheless the Cartesian intuition of distinctness remains very much alive. The conventional view in contemporary philosophy of mind is that although all *things* are materially constituted mental *properties* and *events* are distinct from and *in some sense* irreducible to physical ones. It is not surprising then that the problem of mental causation is still with us now as the question of how mental and physical *properties* and *events* can causally interact.

It would be nice to find a metaphysical framework that is compatible with the causal completeness of the physical sciences, respects the distinctness intuitions, and also provides a place for mental causation. Without the latter it is impossible to make sense of perception, thinking, and action since they inextricably involve causal relations. Non-reductive physicalism (NRP) is claimed to be the metaphysical framework that fits the bill. Proponents of NRP, among which I count myself, endorse physicalism and claim our view to be compatible with the scientific account of the world. We

---

[1] Papineau (2001) contains an excellent discussion of how the idea that physics is causally/nomologically closed (or causally complete) became so persuasive and the problems this poses for various forms of dualism. See also Kim (2005) for the problem of making sense of causal interactions between mental and material substances. A quite different line of criticism of dualism was pursued by philosophical behaviorists (e.g. Ryle 1949) who emphasized what they took to be epistemological and semantic problems with dualism. While these arguments were quite influential fifty years ago they have now been mostly rejected along with the behaviourism they were thought to support.

also claim that NRP is compatible with irreducibility of the mental and that it allows for mental causation.

During the last few decades, Jaegwon Kim has been telling philosophers of mind that when it comes to the mind–body problem 'you can't both eat and have your cake' and that when it comes to mental causation there are 'no free lunches'.[2] He has developed an argument that he calls the 'exclusion' argument that he thinks shows that no version of NRP can properly accommodate mental causation. The 'stark choices', according to Kim, are dualism, reductionism, elimitivism, and combinations of these. Kim's own response to the exclusion argument is a view he characterizes as 'physicalism, or something near enough' (Kim 2005). This view is supposed to save mental causation from the exclusion argument as far as possible. In fact, it ends up being quasi elimitivist since it denies that there are intentional *properties as such* but only species specific or person specific intentional properties.[3] It is only 'near enough' because, to the surprise of those of his readers who think of Kim as a paradigmatic reductive physicalist, it also ends up being dualist and epiphenomenalist with respect to qualia.[4] I think that Kim overreacts to his Exclusion argument. My aim here is to articulate and defend a particular version of NRP against Kim's exclusion argument and to make a proposal for how mental causation—or something near enough—can find a home within this version of NRP.

I. What is 'non-reductive physicalism?'

NRP is a metaphysical view of the mind that claims to reconcile physicalism, the irreducibility of mental properties, and mental causation. The way I understand NRP may not be the same as some other writers since I take the 'physicalism' part more seriously than others.[5] My version of NRP is committed to these claims:

(1) Jackson–Lewis Physicalism: Every positive truth and every truth concerning laws and causation is metaphysically necessitated by truths concerning the spatio-tempo-

---

[2] Kim at first seemed to express sympathy with some version of NRP and the hope that the concept of supervenience could provide the key to formulating it (Kim 1993) but he seems to have soon come to the conclusion that NRP cannot properly handle mental causation (Kim 1993) and has been arguing against it for the past quarter century.

[3] Kim argues that a predicate that is multiply realizable (e.g. a typical functional predicate) does not refer to a genuine property but is reducible instance by instance to whatever physical property realizes it on that particular instance. As we will see this account of reduction is partly motivated by the exclusion argument.

[4] The view is proposed in Kim (1998) and is spelled out more extensively in Kim (2005). With regard to phenomenal consciousness Kim says 'qualia, are not functionalizable, and hence physically irreducible . . . There is a possible world just like this world in all respects except for the fact that in that world qualia are distributed differently, (2006, p.170)". I disagree but don't discuss the claim that the non-functionalizability of qualia predicates/concepts is a good reason for dualism. For discussion see Loar (1997) and Balog (1999).

[5] Davidson's 'anomalous monism' is usually thought of as a version of non-reductive physicalism. But its commitment to physicalism may be rather weak since though Davidson does endorse the causal closure of physics it is not clear that he agrees with the supervenience claim I call 'Jackson-Lewis Physicalism' since he denies that there are 'tight connections' between the mental and the physical.

ral distribution of instantiations of fundamental physical properties and relations and the fundamental physical laws.[6]

(2) Physical Nomological Closure: For every physical event proposition E(t') (except events at the initial condition) and every prior time t the probability at t of E(t') given the physical state of the universe at t and the fundamental laws is the objective probability at t of E(t').

(3) Irreducibility: Some mental properties (events) are real and are not identical to any real physical properties (events).

(4) Mental Causation: Mental properties (events) are causally related (cause) to physical (and other) properties (events).

The first two claims characterize a physicalist world view or what I will call 'Physicalism'. Condition (1) expresses the physicalist idea that 'all God needed to do' to make the universe is to distribute the fundamental physical properties in space/time and make the laws of fundamental physics. All facts about macroscopic objects, their colours and behaviors and facts about people, their thoughts and experiences, and truths about causation and the special sciences and so on are metaphysically entailed by the fundamental physical facts and laws. Condition (2) says that the physical laws are closed and complete in the sense that given the complete fundamental physical state at t and the laws whether or not E occurs at t' or its chance at of its occurring at t' is completely determined. I assume that whatever causation is condition (2) implies the *casual* completeness of physics in that E(t')'s physical causes at t are sufficient to determine its occurrence (or the chances of its occurrence). Condition (2) is a consequence of (1) and it is possible to derive (1) from (2) and some other plausible premises but I separate them since nomological and causal closure will figure importantly in our discussion.[7]

Physicalism comes in two varieties, reductive physicalism (RP) and non–reductive physicalism (NRP).[8] RP claims contrary to (3) that every *real* or as I will say '*genuine*' property (G-property) that has instances in our world (or any physically possible world) is identical to a *physical* G- property. NRP claims that some mental properties are G-properties that are not identical to any *physical* G-properties. If events are, as Kim and I think, instantiations of G-properties then there are mental events that are not identical to physical events but are nonetheless *real*. Mental Causation says that some of

---

[6] Frank Jackson (1998) and David Chalmers (1996) characterize physicalism as the claim that every truth is necessitated by totality of truths in the complete language of ideal fundamental physics and the laws of fundamental physics and a statement to the effect that this is the totality of fundamental truths and laws. (The latter can be avoided by restricting the characterization to positive truths.) They hold additionally that the entailments required by Physicalism are a priori. I do not assume that here. David Lewis (1983) earlier provided a similar characterization of physicalism. There are issues concerning how to define 'fundamental physical property or ideal physics' and whether this account is sufficient for physicalism. (It is surely necessary.)

[7] See Papineau (1993) and Loewer (1995) for arguments from (2) to (1).

[8] Some philosophers call themselves (or are called by others) 'physicalists' because they hold that all things are materially constituted even though they reject Physicalism. Perhaps Davidson (1980) and Searle (1992) are examples.

these mental events cause physical events. It is pretty clear that NRP is committed to causal over determination. Later we will look at how Kim formulates an argument that makes this commitment explicit and attempts to refute it.

The disagreement between advocates of RP and advocates of NRP is an 'in-house' argument among physicalists. Obviously the difference between them depends on what counts as a G-property and what counts as a physical property. To explain how I will employ these notions, I will adopt a framework devised by David Lewis though without all of his metaphysical commitments.[9] Lewis calls the conception on which every predicate corresponds to a property the 'abundant conception' of properties. The abundant properties are, or correspond to, sets of possible individuals. This is a 'thick' notion of property since predicates that differ in meaning may correspond to the same property. For example, 'is a puddle of H_2O' and 'is a puddle of pure water' differ in meaning but correspond to the same property.[10] The G-properties are a subset of the abundant properties. Roughly, the idea is that G-properties are those 'that cut nature at its joints'. Lewis (and I) suppose that nature has many joints. First there are the most fundamental joints. Lewis calls the most fundamental properties 'the perfectly natural properties'. Lewis claims that all truths supervene on the totality throughout space-time of the institutions of perfectly natural properties. Since Lewis thinks Physicalism is true he thinks that all the perfectly natural properties exemplified in our world are physical (—mass, charge, spin, —these are properties that occur in proposals for the most fundamental laws of physics).[11] Non-physicalist philosophers might want to include as perfectly natural properties involving phenomenal consciousness and intentionality.[12]

Second are nature's higher level joints. These correspond to the properties or kinds that occur in laws of special sciences. These are also G-properties.[13] By 'law' I mean a simple true generalization or equation that is counterfactual supporting, projectible,

---

[9] Lewis thinks that any class of possible individuals is a property (the 'abundant' conception) and that certain of these classes are, or correspond to, perfectly natural properties (the 'sparse' conception) and that naturalness comes in degrees. He also holds that the degree of naturalness of a property is a matter of metaphysical necessity, that the perfectly natural properties instantiated at our world are all intrinsic to space-time points (or small regions) except for space-time relations, and that all truths- including the laws- supervene on the distribution of perfectly natural properties. The latter two comprise his doctrine of Humean Supervenience. I make none of these assumptions in this paper.

[10] Davidson, whose view Anomalous Monism, is often thought of as a version of NRP was very skeptical about properties but seems to have given in to talking about properties (Davidson 1993). Needles to say, except in a footnote, the view that predicates with non-analytically connected meanings may correspond to the same property depends on a Fregean-like notion of meaning.

[11] Lewis holds both that physics makes the best estimates of the natural properties and that what properties are natural is a matter of necessity. There is a tension between these commitments.

[12] For example, Chalmers (1996) argues for the view that there are fundamental mental or proto-mental properties linked by laws to physical properties. More recently Chalmers (2001) has suggested that mental features might be the categorical basis of fundamental physical properties.

[13] Armstrong (1978) holds this view concerning universals. Of course exactly what this view comes to depends on what laws there are and what it is to enter into a law or causal relation in 'an appropriate way'. Fodor says that *natural kinds* (i.e. genuine properties) are properties that appear in laws and then explains laws by saying that laws are generalizations connecting *natural kinds*. Well, explanation has to end somewhere.

sufficiently simple, and so on.[14] So the G-properties include the perfectly natural properties and any other properties that are involved in laws. Perhaps there are other properties that should be counted as genuine as well.[15]

In order for there to be an interesting difference between RP and NRP there needs to be a restriction on G properties so that not every construction out of physical G 'properties is itself a G'property, and this restriction needs to be of metaphysical and scientific significance. The distinction between properties that are either fundamental or law involving and those that are not is the distinction that most philosophers have in mind when they speak of some properties as 'real' and others as not. Some plausible candidates for G-properties are: *positive charge, being a gas, mutation rate, episodic memory, and being a monetary exchange*. Plausible candidates for not being G-properties are: *being gruesome, being postmodern,* and *being a gas or a mutation*.[16]

I will make an assumption about how G-properties figure in laws and causation with which, I think, Kim would agree. It is that if F is a G-property that figures in the antecedent of a dynamical law then an instantiation of F is (or corresponds to) an event and this event can be a cause of other events. So if there is some reason to think that a certain property instantiation cannot be a cause then that is reason to think that it cannot figure in a dynamical law. Perhaps the converse that every causal property occurs in a law (or corresponds to a predicate that occurs in a law) is also true. But this won't figure in my discussion.

Here is a bit of terminology and some abbreviations that will be useful. A *mental property*, M-property, is any property that corresponds to a mental, that is intentional or a qualia predicate; for example, ';is thinking about soup', 'feels dizzy'. An MG-property is a mental property that is a G-property. A P-property is any property that is picked out by a kind predicate of a natural science. The natural sciences include physics, chemistry, biology, and so on but not intentional/consciousness psychology. Since a disjunction of kind predicates is not necessarily a kind predicate not every broadly physical property is necessarily a P-property. So every P-property is a G-property but it is left open whether M-properties are G-properties and if so are P-properties. There are two important questions for physicalists about M-properties.

(1) Are some M-properties MG-properties?
(2) Are all MG-properties P-properties?

---

[14] These are the usual criteria for lawhood. Something along the following lines is what I have in mind. If FG is a law and Fa is logically compatible with its being a law then FaGi (G is an appropriate instance of the law) is true and positive instance of FG provide confirmation for further positive instances. Many laws of the special sciences hold *ceteris paribus*.

[15] Some philosophers might want to include certain mental properties as 'genuine' even if they are not fundamental and not law involving. I am not sure that these would be 'genuine' in the same sense but it is of no matter since whether or not there are additional genuine properties makes no difference to my discussion.

[16] Of course the instances of any 'gruesome' property fall under laws and can be causes. The claim is that the gruesome property does not itself occur in a law or ground a causal relation.

RP says that if any M-property is a G-property, then it is a P-property. NRP says there are M properties that are G properties and that are not identical to any P-property. In other words, RP says that all of nature's joints are physical while NRP says that some are mental.

I am not arguing for NRP in this paper but will be content to defend it against Kim's arguments. Still, I want to mention briefly the main reasons for a physicalist to be a non-reductive physicalist. First, as Fodor (1974) observed, the property identities that RP claims to exist failed to be found. Second, many philosophers are persuaded that certain features of mental properties establish that they are not identical to neurophysiological or any other P-properties. The features are (a) multiple realizability, (b) externalism, and (c) the existence of an 'explanatory gap' between physical and mental descriptions.[17] I will just say a few words about the first of these. The idea is that certain mental predicates, for example, 'that London is pretty' are realized by various quite different neurophysiological properties in different people's brains (e.g. an English speaker and a French speaker) and even by different neurophysiological properties in a single person at different times. The reason for this is that these properties are functional properties. Their instances must satisfy a certain functional or causal profile that can be satisfied by many processes that are physically dissimilar.[18] It is plausible that there is no genuine physical property that unites the various instances of thinking that London is pretty.

On its face, it is plausible that some M-properties are G-properties since there are many psychological generalizations containing intentional and phenomenal predicates that satisfy the usual characterizations of 'law'; that is, they are confirmable by instances, support counterfactuals, ground explanations, and so on. What looks and acts like a law should be considered to be a law unless there is a good argument that it can't be one.[19] In fact, Kim has argued that there is a conflict between holding that, on the one hand, mental predicates/properties are multiply realizable, and that on the other hand, that they occur in laws (given the completeness of physical laws).[20] If there were such a

---

[17] Explanatory gap considerations are sometimes taken to motivate not only that qualia are not identical to physical properties but that they don't even metaphysically supervene on them and the physical laws. That is, it is taken to show that physicalism is false. In view of Kim's dualist views concerning qualia I suspect that he would agree with this. Physicalists, of course, disagree. See note 6.

[18] Of course, the instances must have enough physically in common to implement the causal roles associated with the functional property but there may be no predicate of any natural science that is satisfied exactly by those systems that implement those roles.

[19] Some philosophers for various reasons do not consider these generalizations to be bona fide laws and want to reserve the term 'law' for fundamental exceptionless laws of physics. That is their prerogative but it doesn't really affect my discussion of NRP since everyone will agree that there are *ceteris paribus* generalizations in the non-mental special sciences (including biology, geology, and chemistry and so on) that play the role with respect to confirmation, counterfactual support, causation, and explanation characteristic of laws. We can call them 'special-science-laws'. Fodor's point, with which I agree, is that whatever you call this kind of generalization they also occur in psychology and involve M-properties. NRP claims that some of the G-properties that occur in psychological special-science-laws are not identical to any P-properties.

[20] The argument is in Kim (1998). Kim observes that certain disjunctive properties- his example is 'is jade' which is equivalent to 'is jadite or is nephrite' is not projectible since the two realizers are physically different. But even if this is true of 'is jade' it doesn't follow that *no* multiply realized predicate occurs in a law.

conflict then any mental predicate that corresponded to G-property would thereby also correspond to a P-property and NRP would be dead in the water. The exclusion argument wouldn't be needed to show this. But there is no obvious conflict between the completeness of physics and the existence of special science laws involving multiply realizable predicates.[21] So let's see if the exclusion argument is more effective.

II. The supervenience/exclusion argument

Kim's supervenience/exclusion argument has received a lot of discussion, so I will be going over some well (in some cases very *well*) tread ground. Here is how Kim recently formulates the exclusion argument (2004, p. 39)

Let M and M$\star$ be mental properties and m and m be the events of M's instantiation at some location and time t and $m\star$ be the event of M$\star$s instantiation at some place and time $t\star$ and suppose that

(1) m causes $m\star$[22]

In saying that m causes $m\star$ Kim is supposing that m is causally sufficient in the circumstances for $m\star$. Since physicalism holds there will be some physical property P$\star$ whose instantiation $p\star$ at time $t\star$ is such that

(2) $m\star$ has $p\star$ as its supervenience base[23]

Kim argues that (1) and (2) support

(3) m caused $m\star$ by causing $p\star$

Since physicalism holds

(4) m also has a physical supervenience base p.

Kim then appeals to a principle he calls 'Closure' (what we earlier called 'the causal completeness of physics')

Closure: if a physical event has a cause that occurs at t, it has a physical cause that occurs at t.

It follows that

(5) m causes $p\star$ and p causes $p\star$

Since we are assuming NRP, i.e. non identity

(6) M $\neq$ P and so $m \neq p$

At this point in the argument Kim appeals to a principle that he calls 'Exclusion'.

---

[21] See Fodor (1992) and Loewer (2006b) for rebuttals of Kim's argument.

[22] Kim talks of properties being in causal relations and also property instances being in causal relations. He identifies events with property instances so the latter involves event causation.

[23] That is there is some physical fact that is metaphysically sufficient for M$\star$. We can think of P$\star$ as the property of this fact obtaining at some region of space-time. Although Kim doesn't emphasize the point P$\star$ may be enormously complicated and may involve events in a temporal region. It may not be a *genuine* physical property.

Exclusion: No single event can have more than *one sufficient* (my italics) cause occurring at any given time-unless it is a case of causal over determination.

But according to Kim this isn't a case of causal over-determination. By causal over-determination Kim means the kind of case in which there are two shooters each of which kills the victim. That seems right.

(7) p* is not causally over determined by *m* and *p*

It follows that either *p* or m does not cause p*. By closure it must be that

(8) The putative mental cause m is excluded by the physical cause *p*. That is, *p* not *m*, is a cause of *p*.

As Kim observes, Supervenience isn't needed for the argument. The conflict is among M≠P, Closure, and Exclusion. It appears then that the argument works equally well against NRP and against non-physicalist views.[24]

III. The exclusion argument defanged

At this point, it may be useful to remind ourselves what is at stake in the exclusion argument. Various considerations (functionalism, multiple realization, externalism, explanatory gap) make M≠P persuasive. Also there are scientifically compelling reasons to accept Physicalism and Closure. Giving up mental causation is a last resort. So Kim's argument is a paradox. Each of M≠P, Closure, Mental Causation, and Exclusion is plausible but together they are inconsistent. We proponents of NRP accept M≠P, Closure, and Mental Causation, so we have to reject Exclusion.

One response to the Exclusion argument is that it must be unsound since the parallel argument in which P is restricted to fundamental physical properties and M is any multiply realized special science property would show either that M is reducible to fundamental physical properties or that M isn't causal.[25] Kim calls this 'the generalization argument' and attempts to rebut it. I don't intend to get into the details of his reply since my primary response to the argument will be to attack Exclusion directly. However, I do want to discuss one way in which Kim responds to the generalization argument since it provides insight into how he is thinking about causation that will be relevant to my criticisms of Exclusion. The heart of his response involves a distinction he makes between levels and orders and his view that causation is grounded in *causal powers*. At the most fundamental level there are elementary particles and at the next level are certain configurations of elementary particles and so one up a ladder of levels. On Kim's account a property P may bestow certain causal powers on an individual X that are not had by any of X's lower level components. For example, a brain possesses

[24] Actually, as I will argue in the next section the argument doesn't work against NRP but has some bite against non-physicalist emergentism.
[25] The Generalization Argument has been made by Gillett (2001), Hansen (2000), Block (2003), among others.

causal powers in virtue of being a brain that is not possessed by its component neurons and neurons have causal powers that are not possessed by their molecular components and so on down to the most elementary particles. Kim's suggests that some special science properties apply to higher level individuals and bestow causal powers not bestowed by lower level properties. A functional property and its realizer properties apply to individuals at the same level but or of different orders. The functional property is second order since it applies in virtue of an individual possessing some first order property that realizes it. On this picture the functional property doesn't contribute any new causal powers over and above its realizer to the individuals that instantiate them. Since Kim supposes that certain psychological properties (although not consciousness properties) are second order and apply at the same level as brain properties they are subject to the exclusion argument. But those special science properties that apply at higher levels and are first order may contribute novel causal powers and so are not subject to the exclusion argument. I think that is basically the idea.

Kim's attempt to save some macro-properties from the exclusion argument depends on this metaphysics of level and orders and on his thinking of causation in terms of the causal powers properties bestow on individuals. I have a lot of doubts about this metaphysical framework.[26] As we will see it doesn't fit well with the way causation is conceived in fundamental physics. In any case it doesn't really provide an adequate defence against the generalization argument. Many special science properties other than psychological properties are functional properties that apply at the same level as their realizers and so are just as subject to the exclusion argument as psychological properties. Further, it seems that the only properties that escape the exclusion argument, if it is sound, are the microphysical states of isolated systems. For example, being a low pressure system is not identical to any particular arrangement of particles and fields since there are infinitely many distinct micro states that make for a low pressure system but at best is necessarily a disjunction of these states. Nomological closure/completeness implies that whatever causal powers a specific low pressure system has it has in virtue of the fundamental physical laws applied to its fundamental physical state. Speaking the language of 'causal powers' it appears that being a low pressure system doesn't contribute any *new* causal powers over and above the causal powers of the microphysical state. Anyway, that's why I don't find Kim's reply to the generalization argument effective. But, as Kim points out, the generalization argument at best shows that *something* is wrong with the exclusion argument (or that it threatens a lot more than we might have thought) it doesn't tell us *what* is wrong explain *why* it is wrong. A better response to the exclusion argument is to show why Exclusion is wrong. So let's examine this premise more carefully.

---

[26] I am not at all sure how to assign levels to configurations of particles. Are water molecules and sugar molecules at the same or different or incomparable levels? What if fundamental physics says (as it likely does) that what basically exists are fields of various kinds? Kim discusses some of the problems with 'the layered' account in Kim (2002).

Exclusion says that 'no single event can have more than one sufficient cause occurring at any given time—unless it is a genuine case of causal over determination'. By 'a genuine case of causal over-determination' Kim has in mind the type of situation in which two assassins fire simultaneously at the victim causing her death. In this kind of situation the two events (the two firings) are *metaphysically* independent and each involves its own causal process that culminates in the death of the victim. Causal over determination like this may be rare but it is not metaphysically problematic. Kim observes that the putative situation in which a non-physical genuine property instantiation $M(y,t)$ and its physical realizer $P(y,t)$ are said to both cause $Q(z,t')$ is not a case of genuine causal over determination like this. He is certainly correct about this since $P(y, t)$ and $M(y,t)$ are not metaphysically independent. Let's call the kind of over determination involved in mental/physical causation 'M-over determination' and understand Kim as ruling it out by the exclusion principle. According to exclusion the putative mental and physical causes of $Q(z,t')$ compete and so one is not really a cause of $Q(z,t')$. Since NRP assumes that the physical realm is causally closed $P(y,t)$ wins the competition and $M(y,t)$ is not a cause of $Q(z,t')$.

The exclusion argument assumes that the physical cause and the putative mental cause are *sufficient* causes of $Q(z,t')$. However, if $M(y,t)$ and $P(y,t)$ are ordinary macro events then, contrary to Kim's supposition, they are certainly not by themselves sufficient for $Q(z,t')$. It is common place among philosophers of science, but perhaps not as recognized as it should be outside of philosophy of science, that for any small region R of space at time t nothing much short of the state of the universe in a sphere with center R and whose radius is one light second (i.e. 186,000 miles) at t-1 second is causally sufficient for determining what will occur (or the chances at t-1 of what will occur) in R.[27] Because of this, I suggest that we interpret Kim's exclusion principle not as involving causes that are literally sufficient for their effects but as nomologically sufficient in the circumstances C where C is a partial description of the state at t.

Exclusion*: there can't be two distinct events $P(x,t)$ and $F(x,t)$ such that both are causes of $Q(z,t')$and there are circumstances C such that both $P(x,t)$ and $F(x,t)$ are nomologically sufficient for $Q(z,t')$ in C for $Q(z,t')$ (or are each nomologically sufficient in C for the chance at t of $Q(z,t')$).

Kim doesn't argue for Exclusion or Exclusion* since he thinks that it 'virtually an analytic truth without much content' (p. 51). This is puzzling since it is sufficiently contentful to play an essential role in the argument that NRP is incoherent and philosophers who deny it, myself among them, don't think that we are denying an analytic truth. So there must be some assumptions about the nature of causation that

---

[27] Or if the fundamental laws are deterministic as determining the chance at t of $Q(z,t')$. More exactly, for any event E at t' there will be a physical proposition K that holds at time t that is a minimal sufficient condition for the occurrence of E (given the physical laws) which is typically a partial description of the complete state at t (or state on a hyper surface intersecting t) but this proposition will involve values of physical parameters throughout the hyper surface. This point is made by Latham, Field, Loewer, and Elga among others.

Kim accepts and that lead him to think that M-over-determination is 'virtually analytic'. I don't know what exactly he is assuming but I want to offer a speculative hypothesis.

Ned Hall (2004) has recently argued that there are two concepts of causation which he calls 'production' and 'dependence'. Production is the relation that supposedly obtains when one billiard ball hits another and thus *produces* motion in the second. Dependence is the relation that holds between two events when features of the second (including whether or not it occurs) counterfactually depends on features of the first (including whether or not it occurs). Hall thinks that the two kinds often go together but are fundamentally different and that it is possible to have one without the other. For example, the kitchen fire may depend on my forgetting to turn off the heat under a pot but my forgetting does not *produce* the fire. Billy's throw may produce the breaking window even though, in this instance, *dependence* is absent since Billy's rock arrived a moment before Sally's which would have broken the window anyway.

My diagnosis of why Kim thinks that exclusion is virtually analytic is that he is thinking of causation as *production*. If one thinks about causation in this way then it is quite natural to see exclusion as virtually analytic. If $P(x,t)$ literally produces $Q(z,t')$ then it does appear that 'there is no work' left for any other event $F(x,t)$ to do as far as producing $Q(z,t')$. Kim also seems to think of the causal relation as involving the transfer of some quantity, *causal oomph*, from the cause that brings the effect into existence. It is not surprising then that he would think that a second dose of *oomph* from $F(x,t)$ is not only not needed to produce $Q(z,t')$ there isn't even any place for it.

I am not confident that Kim is thinking of causation in this way but I am sure that no one who accepts Physicalism should endorse it. In a famous essay, Bertrand Russell noted that causation as a relation of production connecting local events makes no appearance in the ontology of the fundamental laws of physics and suggested that the idea of causation '... is a relic of a bygone age, surviving, like the monarchy, only because it is erroneously supposed to do no harm'.[28] The problem he finds with it is that while productive causation connects localized events at different times the candidates for fundamental laws are differential equations that connect global states with one another. If one wants to speak of 'production' it seems that it is the whole state and the laws that produce subsequent states. Russell's suggestion that we would be better off without the concept of causation, as he also came to believe, is an overreaction. The appropriate response for a physicalist is to characterize causal relations that supervene on the fundamental laws and facts. I don't know whether there is account of causation as production that can support exclusion and is also compatible with Physicalism. But

[28] 'All philosophers, of every school, imagine that *causation* is one of the fundamental axioms or postulates of science, yet, oddly enough, in advanced sciences such as gravitational astronomy, the word 'cause' never appears. Dr. James Ward[...]makes this a ground of complaint against physics[...]To me, it seems that[...] the reason why physics has ceased to look for causes is that, in fact, there are no such things.[...]The law of causality, I believe, like much that passes muster among philosophers, is a relic of a bygone age, surviving, like the monarchy, only because it is erroneously supposed to do no harm'.

even if there is, all it would establish is that NRP is not compatible with mental events being productive causes so understood. This conclusion would have the import that Kim thinks his exclusion argument has only if it is causation as production (construed in this way) that we think is really required to vindicate mental causation. But I think a good case can be made that causation as dependence will do perfectly well.

Causation as dependence

Counterfactuals are notoriously vague and context dependent. The way they should be understood for the purposes of characterizing causation as dependence is along the lines (but not quite; see note 32) of David Lewis' famous account. On that account A>B is true if either there are worlds at which A&B are true that are more similar to the actual world than any world at which A&-B is true. Lewis specifies a particular account of world similarity that he thinks has the consequence that in evaluating A>B one looks at worlds that are identical to the actual world from the worlds initial condition and then diverges from the actual world (perhaps this requires a violation of actual laws in a small region for a short time) and then evolves in accord with the laws of the actual world so that A is true. If all these worlds are all worlds at which B is true, then A>B is true. For example, 'if at noon Terry had wanted a beer he would have opened the refrigerator' is true if the worlds that are identical to the actual world up until a moment prior to noon when a small miracle occurs so that Terry is in a brain state that realizes wanting a beer are also worlds where he opens the refrigerator.[29]

Lewis says that E depends on C iff C and E are non-overlapping events and if C had not occurred the E would not have occurred. His original account of causation was that C causes E iff C and E occur and E depends on C or there is a chain of events connected by dependence from C to E. This account is vulnerable to cases of preemption in which C causes E but E doesn't depend on C because there is another event C* waiting in the wings to cause E if C didn't occur. Lewis modified his account in a way the handles many pre-emption counter-examples. On the most recent account, an event E *influences* an event C if E and C don't overlap and if there are suitable variations in C that are counterfactually correlated with variations in E. C causes (in the dependence sense) E iff C and E occur and there is a chain of events connected by influence from C to E. For example, the height of mercury in a thermometer depends on the ambient temperature since the counterfactuals 'if the temperature had had been (or were to be)

---

[29] Unfortunately, Lewis's account of world similarity doesn't have the consequence he thinks it has. The heart of the problem is that his account of similarity involves laws and other considerations that are temporally symmetrical while the similarity he thinks the gets out of these considerations is temporally asymmetric, as it must be if it is going to get the truth values connected with causation as dependence correct since these counterfactuals are temporally asymmetric. See Elga (2001). Jonathan Bennett (2003) characterizes truth conditions of counterfactuals by simply counting past perfect match and not future march as making for similarity. It is possible to fix this all up by adding a bit to Lewis account so that one gets more or less the similarity relation Lewis was aiming at but it would take us to far a field to do it here. See Loewer (2006) for the fix up.

x, the height of the thermometer would have been (would be) y' are true for a range of x and y.

It is plausible that under normal conditions small differences in a person's brain corresponding to different mental states (e.g. different intentions) lead by law to correspondingly different bodily movements. That is, that counterfactual dependencies *on Lewis' construal of counterfactuals* between mental events and bodily events obtains. If so then mental events cause *in the dependence sense* bodily events. My proposal is not that Lewis' influence account perfectly captures our intuitive concept of causation. But I do claim that causation as influence is near enough to our folk conception of mental causation to underwrite the role of causation in folk psychology, rational deliberation, action theory and so on. In the remainder of this paper I will lay out a case for this claim.

The first thing to note is that there is no problem of over-determination if causation is understood as dependence. On Lewis' account of counterfactuals, a particular event (or the value of a range of possible events) can depend on many co-occurring events. The motions of one's body, for example, the motions of a person's arms and hands when reaching into the refrigerator depends counterfactually both on her mental states (which snacks she wants) and on her brain (and other bodily) states, and on a myriad of other states and events. Also, the kind of 'M-over determination' involved in B depending on both M and P is neither like the two assassins kind nor the production kind. In particular there is no temptation to say that if B depends on P it can't also depend on M since 'there is no work for M to do'. If there is 'work being done' it is being done by the *fundamental* dynamical laws that evolve the entire state. The influence counterfactuals connect aspects of the state at one time to aspects at other times so that alterations of one correlate with alternation of the other.

Kim expresses his worries about counterfactual accounts in this passage:

To summarize our discussion of the counterfactual approach then, what the counterfactual theorists need to do is to give an *account* of just what makes those mind-body counterfactuals we want for mental causation true and show that on that account those counterfactuals we don't want, for example epiphenomenalist counterfactuals, turn out to be false. Merely to point to the apparent truth, and acceptability of certain mind-body counterfactuals as a vindication of mind body causation is to misconstrue the philosophical task at hand[. . .]what we want—at least what some of us are looking for—is a philosophical account of *how* it (mental causation and the corresponding counterfactuals) can be real in light of other principles and truths that seem to be forced upon us.

I agree with Kim that merely pointing out that certain counterfactuals are true or appear to be true is not sufficient to ground mental causation. As he says, it is also required for there to be an account of how those counterfactuals come to be true and further that account cannot presuppose mental causation and must be compatible with NRP. Lewis' account does this. On Lewis, account the fundamental laws and facts of physics is 'what makes those mind-body counterfactuals we want for mental causation' true.[30] 'If I were to decide to get a beer I would walk over to the refrigerator' (and similarly for the battery of other counterfactuals that ground causation as dependence) is true when the worlds most similar to the actual world in which I decide to get a beer are worlds in which I walk over to the refrigerator. Whether that is so depends on the actual laws of physics (since what they are determines what counts as a 'small violation') and on the actual physical facts. It is clear that this account of counterfactuals does not presuppose causation or mental causation and is compatible with physicalism. Also, we have seen that there is no problem about over-determination, so the account is compatible with $M \neq P$, i.e. with NRP. So the issue that remains whether the account of counterfactuals really underwrites 'those counterfactuals we want' and not 'those we don't want'. Fully establishing these claims is not something that I can do since it would involve establishing the truth values of many counterfactuals (on Lewis' construal) and that can literally be done only by knowing the physical realizers of mental states and the fundamental laws. But I think I can go some distance towards making the claim plausible and replying to Kim's arguments that causation as dependence cannot do the work we want mental causation to do.

Kim suggests that there may be dependence/influence where there is no mind-body causation. If so then dependence is too weak to ground genuine mind-body causation. He mentions four kinds of situations: backtrackers, common causes, epiphenomenalism, and omissions, where dependence holds but there is no causation.

Although Kim doesn't discuss backtracking in detail I think he may have something like the following worry in mind. On Lewis' account the worlds relevant to evaluating $A(t)>B$ are ones in which at the nearest time prior to t there is a small violation of law that leads to $A(t)$. But for some $A(t)$ that time may be much prior to t. For example, had Haley's comet intersected the orbit of Jupiter at t it would have had to intersect the orbit of Saturn at t-k. Getting the comet intersecting Jupiter's orbit at t when in fact the comet is near Saturn at t would involve too big a miracle at times after t-k.[31] For the planetary system it may be that the past counterfactually depends on the present as much as the future does. This is part of the reason that I don't think Lewis' account by itself corresponds exactly to our notion of causation. One would have to add some further conditions (perhaps that under usual conditions causation is counterfactual dependence from past to future). But this is not a problem for mental causation because various decisions, intentions, and so on correspond to very small differences in the brain. Partly for this reason it will always be possible for the violation of law that is

---

[30] Kim suggests (2006) that the truth makers of counterfactuals or the counterfactuals that go along with mental causation involve causation as production. This is correct if one has in mind the fundamental physical laws evolving fundamental physical states. But Kim is more likely thinking of what I called 'local production'. Relations of local production are not the truth makers of counterfactuals on Lewis's account. The fundamental laws and fundamental physical state are the ultimate truth makes of both kinds of causal relations.

[31] Note that the point isn't that we are evaluating the counterfactual in what Lewis thinks of as a backtracking way of evaluating similarity but rather that Lewis' way of evaluating similarity leads to back tracking in certain situations.

required for a counterfactual mental event to occur almost immediately prior to the mental event. I will return to this point later.

C is a common cause of A and B when C causes both A and B but there is no causal relation between A and B. For example, a rock thrown into the centre of a pool (C) causes a wave to hit at point *a* and at point *b* at time t. The worry is that the counterfactual 'if A had not occurred then B would not have occurred' may appear to be true. In fact, I think that in ordinary language this counterfactual is plausibly true in the situation I described. But recall that the characterization of causation as dependence involves a very particular way of evaluating counterfactuals. On that way, this counterfactual is false since the world in which a small violation of law occurs just before t that leads to A not occurring but leaves everything else the same including B is more similar world to the actual world than the world that also leads to the wave not hitting *b* at t. Again, as in the backtracking case, there may be systems that are so set up so that one does obtain counterfactual dependence between events that are effects of a common cause. But this won't occur with respect to mental events and their putative effects.

An interesting example of a possible common cause situation has come up in the philosophy of mind concerning the relationship between the conscious decision to act in a certain way. There is evidence that at least in some cases the decision and act are related as the common effects of a brain event that is the common cause of both.[32] Whether or not this is so it is clear that causation as dependence/influence has no trouble distinguishing between the decision being the cause of the act or the two being common causes of an unconscious brain event.

Kim raises another worry about dependence that is related to the common cause objection. He argues that causal dependence cannot distinguish the situation in which mental events are genuine causes from the view in which they are mere epiphenomena that are nomologically correlated with brain events that are the genuine causes. Kim pictures the situation involving mental causation as follows in his favourite diagram:

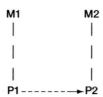

An epiphenomenalist like T. H. Huxely holds that P1 and P2 are events that are sufficient—in senses to be specified—respectively for the events M1 and M2 and that there is a genuine causal relation between P1 and P2 but not between M1 and P2 (or M2). Kim claims that this is completely compatible with both P1 and M1 causing in

---

[32] Wegner (2002).

the dependence P2. How one should respond to Kim depends on how he is thinking of the strength of the relation between the Ps and the Ms depicted by the *vertical* lines. Epiphenomenalists generally think of this relation as *weaker* than metaphysical necessitation. Perhaps it is nomological.[33] Kim likes to illustrate epiphenomenalism with the example of the positions of a shadow cast be a moving ball that *seem* to be causally connected. The positions of the shadow are nomologically connected to the positions of the ball that casts the shadow but are not causally related to each other. Kim seems to think that the counterfactual account fails to count this as epiphenomenalism since M1>-P2 will be true. But if counterfactuals are evaluated along Lewisian lines it is not clear that we obtain this result. It is plausible that −M1>-P2 fails since the most similar world in which −M1 holds is one in which the vertical law connecting P1 to M1 is broken while the horizontal law connecting P1 to P2 continues to hold.[34] On the other hand, -P1>-P2 may be true. In contrast to this NRP holds that the connection between P1 and M1 is one of metaphysical not merely nomological necessitation. In the most similar worlds at which −M1 it is also −P1 since there is no question of 'breaking' the metaphysical connection. So in this situation −M1>-P2 may well be true. But it would be question begging to say that M1 isn't *really* a cause P2 in this case say because it doesn't produce or transfer *oomph* to P2.

The last problem is that Kim points out that dependence can connect omissions with events. Kim says:

> Friends of the counterfactual approach often tout its ability to handle omissions and absences as causes and the productive/generative approach's inability to account for them. We are inclined to take the truth of a counterfactual like:
> If Mary had watered my plants, the plants would not have died as showing that Mary's not watering, an omission, caused the plants' death and take that as a basis for blaming Mary for killing the plants. But obviously there was no flow of energy from Mary to the plants during my absence (that exactly was the problem!); nor was there any other physical connection, or any spatiotemporally contiguous chain of causally connected events.

Kim's objection seems to be that since dependence can connect an omission (Mary's not watering the plants) with an event (the plant's dying) even though there is no transfer of energy from Mary to the plants dependence cannot really be what we want by mental causation. He says of it 'this is not causation worth having'. But, in the first place, unless Mary is outside of the back light cone of the plant's death there will almost certainly be some energy transferred from her to the plants, just not in the right way to save the plants. In any case, since *omissions are not events it doesn't follow from* the fact that there is dependence on omissions that dependence on commissions, and specifically the

---

[33] Chalmers (1996) suggests that qualia are connected by law to physical systems and are epiphenomenal.

[34] Of course there are contexts in which the counterfactual 'if the shadow had not been at position x at time t the ball would not have been at position y a time t+' but it is important to keep in mind that the relevant account is Lewis'. On that account the counterfactual comes is false since small violations in law that changes the position of the shadow leave the position of the ball as it was.

counterfactual sensitivity of the positions of one's body (and, fingers, and so on) to one's volitions and the counterfactual sensitivity of one's volitions on ones intentions, beliefs (and so on) is 'not causation worth having'. Indeed, these relations of dependence and influence are absolutely essential to mentality and action. If the transfer of energy is involved in any case of genuine mental causation, it is also likely involved in any case of mental causation as influence. But the mere transfer of energy certainly isn't what we want for mental causation! My conclusion is that Kim has not shown that counterfactual dependence (underwritten by an account of counterfactuals along Lewisian lines) is not sufficient for genuine mental causation.

I want to conclude with a few sketchy and perhaps surprising remarks about the connection between Lewisian accounts of counterfactuals and mental causation. I have appealed to Lewis' account of causation as dependence to ground mental causation. But there is a way in which mental causation, or more precisely, our neural/cognitive structure, also grounds Lewis' account of counterfactuals and thus causation as dependence.

On Lewis' account the candidates for most similar worlds in which the counterfactual antecedent A(t) is true are those whose pasts match the actual past until a short (or as short as can be) time prior to t and then diverges by a small local violation of law and then evolves in accordance with the actual laws. But why, we may ask are we interested in *this* notion of similarity among the infinity of possible similarity relations that satisfy Lewis semantics for conditionals?[35] One might think that the answer is that this relation is or at any rate is close to tracking the causal relation and we are interested in that relation because it is a fundamental relation between events. But I think this has things backwards. My view is that we are interested in the causal relation not because it is a fundamental relation—there is no fundamental causal relation to be found in physics that connects local events in the way causation is alleged to—but rather because Lewis' account tracks our ability to influence the likelihoods of events. Here is what I have in mind. We assume that the alternative decisions that we might make in the next few moments correspond to very small local physical differences in our brains. That is, different decisions that one might make are realized in differential brain phenomena that can result via the laws from tiny microscopic differences immediately prior to the decisions. If the laws are deterministic, then these small differences from actuality involve small localized violations of law. If the laws are indeterministic, then the alternative decisions can be reached by chance.

Naturally, we are interested in what will happen on the alternative hypotheses of each decision. Of course, that depends not only on the decision but also on many other matters in the environment. For example, suppose Nixon is deciding whether to press the button marked marked 'Launch'. Assuming that Nixon's body, hand, fingers and so on are appropriately connected to his brain then what will happen depends on the buttons being hooked up to a various further devices. *The interesting point for us is that*

[35] This question is asked by Horwich (1987) and answered more fully than by suggestion here in Loewer (2006).

what will happen, or if we allow probabilities over micro histories, the probabilities of what will happen are given by adding one or the other decision to the state that is most similar to the actual state that contains the brain state corresponding to the decision. So the reason we are interested in evaluating counterfactuals along Lewisian lines (or rather along the lines that he thought his proposal yields, and my amended account does yield as mentioned in footnote 30) is that conditionals so evaluated contain information about the likely results of our decisions and this information is enormously important to our getting what we want. If this is on the right track, then causation as dependence has its origin and is most at home in 'the actions of our will;' that is in mental causation. Counter-factual dependence evaluated in terms of the Lewisian account of similarity (and my probabilistic account) can be seen as a generalization from the decision situation. The worlds that count as most similar are those that match the actual world until a short (or as short as possible) time before the antecedent and then lead by a tiny local violation of law (of the sort required for alternative decisions) to the antecedent. Causation as dependence is then characterized in terms of this counterfactual. And causation as dependence *is* causation, or 'something near enough' to be genuine mental causation.

Acknowledgments

I am very grateful to Louise Antony, Karen Bennett, Tim Crane, Carsten Hansen, Brian Mclaughlin, and especially Jaegwon Kim for comments on earlier versions of this paper and for years of discussions. While this paper is a criticism of Jaegwon's views on mental causation, there is no other work on the subject that I admire more or from which I have learned more.

References

Armstrong, D. (1978) *Universals and Scientific Realism*. Cambridge: Cambridge University Press.
Balog, K. (forthcoming) "Mental Quotation."
Balog, K. (1999) Conceivability, Possibility, and the Mind-Body Problem. *Philosophical Review* 108/4: 497–528.
Bennett, J. (2001) *A Philosophical Guide to Conditionals*. Oxford: Oxford University Press.
Bennett, K. (2003) Why the Exclusion Problem Seems Intractable, and How, Just Maybe, To Tract It. *Noûs* 37/3: 471–497.
Bennett, K. (2008) Exclusion Again. In Jakob Hohwy and Jesper Kallestrup (eds), *Being Reduced*. Oxford: Oxford University Press, pp. 280–306.
Block, N. (1994) Functionalism. In S. Guttmplan (ed.), *Blackwell's Companion to the Philosophy of Mind*. Oxford: Blackwell, pp. 317–332.
Block, N. (2003) Do Casual Powers Drain Away? *Philosophy and Phenomenological Research* 67: 133–150.
Chalmers, D. (1996) *The Conscious Mind*. Oxford: Oxford University Press.
Crane, T. (1995) The Mental Causation Debate. *Proceedings of the Aristotelian Society, Supplementary Volume* 69: 211–236.

Crisp, T. and Warfield, T. (2001) Kim's Master Argument. *Noûs* 35: 304–316.

Davidson, Donald (1980) Mental Events. Reprinted in his *Essays on Actions and Events* Oxford and New York: Oxford University Press, pp. 207–224.

Elga, A. (2001) Statistical Mechanics and the Asymmetry of Counterfactual Dependence. *Philosophy of Science* 68 Proceedings, 313–324.

Elga, A. (2006) Isolation and Folk Causation, forthcoming in *Russell's Republic: The Place of Causation in the Constitution of Reality*, ed. Huw Price and Richard Corry. Oxford: Oxford University Press, pp. 106–119.

Field, H. (2003) Causation in a Physical World. In M. Loux and D. Zimmerman (eds), *Oxford Handbook of Metaphysics*. Oxford: Oxford University Press.

Fodor, J. (1974) *The Language of Thought*. Cambridge, MA MIT Press.

Fodor, J. (1992) *Psychosemantics*. Cambridge, MA: MIT Press.

Gillett, C. (2001) Does the Argument from Generalization Generalize? Responses to Kim. *Southern Journal of Philosophy* 39: 2001.

Gillett, C. and B. Loewer (2001) *Physicalism and its Discontents*. Cambridge: Cambridge University Press.

Hall, N. (2004) Two Concepts of Causation. In J. Collins, N. Hall, and L. Paul (eds), *Causation and Counterfactuals*. Cambridge, MA: MIT Press, pp. 225–276.

Hansen, C. (2000) Between a Rock and a Hard Place. *Inquiry: An Interdisciplinary Journal of Philosophy* 43/4: 451– 491.

Horgan, T. (1993) Nonreductive Physicalism and the Explanatory Autonomy of Psychology. In S. Wagner and R. Warner (eds), *Naturalism: A Critical Appraisal*. Notre Dame, IN: Notre Dame University Press, pp. 295–320.

Horwich, P. (1987) *Asymmetries in Time*. Cambridge, MA: MIT Press.

Jackson, F. (1998) *From Metaphysics to Ethics: A Defense of Conceptual Analysis*. Oxford: Oxford University Press.

Kim, J. (1993) *Supervenience and Mind*. Cambridge: Cambridge University Press.

Kim, J. (1998) *Mind in a Physical World*. Cambridge: Cambridge University Press.

Kim, J. (2002) The Layered Model: Metaphysical Considerations. *Philosophical Explorations* 5: 2–20.

Kim, J. (2005) *Physicalism: Or Something Near Enough*. Princeton University Press.

Kim, J. (2006) Causation and Mental Causation. In Brian P. McLaughlin and Jonathan D. Cohen (eds), *Contemporary Debates in Philosophy of Mind*. Oxford: Blackwell, pp. 227–242.

Lewis, D. (1983) *Collected Papers*, Vol 1. Oxford: Oxford University Press.

Lewis, D. (1986) *Collected Papers*, Vol. 2. Oxford: Oxford University Press.

Lepore, E. and B. Loewer (1987) Making Mind Matter, Chapter 11.

Loar, B. (1997) Phenomenal States. In Block, Flanagan, and Guzeldier (eds), *The Nature of Consciousness*. Cambridge, MA: MIT Press, pp. 597–612.

Loewer, B. (1995) An Argument for Strong Supervenience. In E. Savellos and Ü. D. Yalçin (eds), *Supervenience: New Essays*. Cambridge, MA: Cambridge University Press, pp. 218–225.

Loewer, B. (2001a) From Physics to Physicalism. In Gillett and Loewer, pp. 37–57.

Loewer, B. (2001b) Review of Jaegwon Kim's *Mind in a Physical World*. *The Journal of Philosophy* 98: 315–324.

Loewer, B. (2002) Comments on Jaegwon Kim's *Mind in a Physical World*. *Philosophy and Phenomenological Research* 65: 655–662.

Loewer, B. (2006) Counterfactuals and the Second Law. In Huw Price and Richard Corry (eds), *Causal Republicanism*. Oxford: Oxford University Press.

Loewer, B. (2009) Why There is Anything Except Physics. *Synthese* 170/2: 217–233.

Mclaughlin B. (1989) Type Epiphenomenalism, Type Dualism, and the Causal Priority of the Physical. *Philosophical Perspectives* 3: 109–134.

Melnyk, A. (2003) *A Physicalist Manifesto*. Cambridge: Cambridge University Press.

Melnyk, A. (2005) Jaegwon Kim's *Physicalism, or Something Near Enough, Notre Dame Philosophical Reviews* 2005: 7–17.

O'Connor, T. and H. Wong (2005) The Metaphysics of Emergence. *Noûs* 39: 658–678.

Papineau, D. (1993) *Philosophical Naturalism*. Oxford: Blackwell.

Papineau, D. (2001) The Rise of Physicalism. In Gillett and Loewer, pp. 3–37.

Putnam, H. (1975) *Philosophical Papers*, Vol. 2. Cambridge: Cambridge University Press.

Russell, B. (1913) On the Notion of Cause. *Proceedings of the Aristotelian Society* 13: 1–26.

Ryle, G. (1949) *The Concept of Mind*. Chicago: University of Chicago Press.

Seale, J. (1992) *The Rediscovery of the Mind*. Cambridge, MA: MIT Press.

Shoemaker, S. (2001) Realization and Mental Causation. In Gillett and Loewer, pp. 74–99.

Tooley, M. (1987) *Causation: A Realist Approach*. Oxford: Clarendon Press.

Wegner, D. M. (2002) *The Illusion of Conscious Will*. Cambridge, MA: MIT Press.

Index of Names

Lakoff, G. 31
Larson, R. and Segal, G. 115
Leeds, S. 112, 156
Lewis, D. 9, 11–13, 31–2, 35–6, 47,
    119–20, 151, 156, 172, 175–6, 196–7,
    199, 201–2, 209–10, 216, 218, 226–8,
    230–2
Loar, B. 59, 93, 107, 141
Locke, J. 132 n. 2
Lycan, W. 59, 62, 74, 76, 107, 126

Mackie, J. 159, 168 n. 1
McCawley, J. 31
McDowell, J. 76, 117
McGinn, C. 59, 60, 62, 64–8, 71–3, 75–6,
    201 n. 13, 208
McLaughlin, B. 169 n. 6, 189–90
Montague, R. 31–2, 35–8, 41, 43–4

Nozick, R. 122

Papineau, D. 197, 208, 215 n. 1, 217 n. 7
Partee, B. 31–2, 36, 38, 41, 43, 115, 119
Peacocke, C. 208
Perry, J. 61–2, 74
Platts, M. 21–3, 90, 109
Potts, T. 36
Putnam, H. 28 n. 8, 54–5, 59–65, 67, 135,
    138–40, 146, 148, 151–65, 179, 184,
    195, 212

Quine, W. 103 n.5, 104, 115, 122

Rey, G, 206
Richard, M. 121
Russell, B. 225

Salmon, N. 110
Schiffer, S. 100 n. 4, 107, 109–10, 112,
    115–18, 125–6, 128
Searle, J. 134, 138, 140, 144, 217 n. 8
Segal, G. 115, 117 n. 5, 122
Sellers, W. 65–6, 74
Shoemaker, S. 196 n. 4
Soames, S. 115
Sosa, E. 100 n. 4, 169–71, 173–5
Stalnaker, R. 32, 110, 159, 172
Stich, S. 115, 146
Suppes, P, 51

Tarski, A. 19–20, 23, 28, 39, 41, 43, 65, 88–9
Taylor, B. 79 n. 3
Tennat, N. 29
Thomason, R. 38, 41–2, 44, 115

van Fraasen, B. 35, 152
van Inwagen, P. 155
Vernazen, B. 31, 119

Wallace, J. 93
Weinberg, S. 210
White, S. 61, 143 n. 13, 144
Witmer, S. 196 n. 3, 198
Wright, C. 115

Zemach, E. 61

Subject Index